The Perception and Cognition of Visual Space

Paul Linton

The Perception and Cognition of Visual Space

palgrave
macmillan

Paul Linton
Centre for Applied Vision Research
  City, University of London
London, UK

ISBN 978-3-319-66292-3   ISBN 978-3-319-66293-0   (eBook)
DOI 10.1007/978-3-319-66293-0

Library of Congress Control Number: 2017949454

Cover illustration: © Stephen Bonk/Fotolia.co.uk

Printed on acid-free paper

This Palgrave Macmillan imprint is published by Springer Nature
The registered company is Springer International Publishing AG
The registered company address is: Gewerbestrasse 11, 6330 Cham, Switzerland

PREFACE

This book argues for a complete revision in the way we think about the distinction between the perception and the cognition of visual space. For too long the literature has offered a false dichotomy between (a) the perception of visual space and (b) our conscious judgements of visual space. In this book I argue that there is a distinct intermediate level, namely (c) our unconscious post-perceptual judgements of visual space. I argue that pictorial cues operate at this level, and consequently so too must Cue Integration (the integration of depth cues into a single coherent percept). Not only does this argument require us to re-evaluate (a) the distinction between perception and cognition, (b) Cue Integration, and (c) pictorial space, it also requires us to reconsider (d) how we experience depth with one eye. It also has implications for neurophysiology (if Cue Integration operates at the level of V2, does this imply V2 is engaged in cognition rather than perception?) and clinical psychology (questioning the thesis that Cue Integration is reduced in schizophrenia because subjects see 'more clearly through psychosis'). A full introduction is provided by Chap. 1. This book began as a response to the literature on depth from one eye, and I am very grateful for the comments and encouragement that I received from Jan Koenderink and Dhanraj Vishwanath on this initial draft in September 2015, as well as to Chris Frith and Vid Simoniti for kindly reading over that work. I also benefitted immensely from the comments of two anonymous reviewers, a discussion with Brian Rogers, and the presentation of elements of this work at City, University of London in April 2016 to Simon Grant, Michael Morgan, Joshua Solomon, and Christopher Tyler. Since April 2016, this work has continued under the supervision of Christopher Tyler whose knowledge, insight, and support have been absolutely invaluable and have improved this work immeasurably. I cannot thank him enough. I must also thank Dhanraj Vishwanath for his extensive and invaluable comments on the whole manuscript in September 2016 and his further comments in December 2016 and March 2017, and Marty Banks for his extremely helpful and constructive comments in September 2016 on Cue Integration in Chap. 2 and defocus blur in Chap. 4. I must thank Jan Koenderink for very kindly providing me with illustrative examples of his Paradoxical Pseudo-Parallax principle (Koenderink, 2015), which he has kindly permitted me to reproduce in Fig. 3.7 (© Jan Koenderink, all rights remain with him). I must also thank Natasha Reid for preparing Figs. 4.2 and 4.3, and Hikaru Nissanke for preparing the illustrations for an earlier draft. Finally, I must express my gratitude to all of the Psychology team at Palgrave: Laura Aldridge, Sarah Busby, Eleanor Christie, Zaenab Khan, Joanna O'Neill, and Sharla Plant.

June 2017   Paul Linton

The original version of the book was revised: Post-publication corrections have been incorporated. The erratum to the book is available at https://doi.org/10.1007/978-3-319-66293-0_5

CONTENTS

LIST OF FIGURES

3 Stereopsis in the Absence of Binocular Disparity

4 The Physiology and Optics of Monocular Stereopsis

LIST OF TABLES

CHAPTER 1

Two Conceptions of Stereopsis

Abstract In this introductory chapter, I outline the two competing conceptions of stereopsis (or depth perception) that have dominated the literature over the last 150 years. The first conceives of stereopsis in purely optical terms, typically as an exercise in inverse optics. By contrast, the second approach argues that optical information from the world is indeterminate until contextual meaning has first been attributed to it. In this book I attempt to advance a purely optical account of stereopsis and I use this introductory chapter to raise the central contention of Chaps. 2 and 3, namely that many of the 'perceptual' phenomena that appear to count against a purely optical account of stereopsis are better understood as post-perceptual cognitive inferences.

Keywords Stereopsis · Visual cognition · Cue integration
Gestalt psychology · Intentionality

If vision is concerned with the perception of objects (Gibson 1950; Strawson 1979), then stereopsis is the visual space in which those visual objects are located, specifically: (a) the volume of space which each object takes up (in

---

The original version of this chapter was revised: Post-publication corrections have been incorporated. The erratum to this chapter is available at
https://doi.org/10.1007/978-3-319-66293-0_5

© The Author(s) 2017
P. Linton, *The Perception and Cognition of Visual Space*,
DOI 10.1007/978-3-319-66293-0_1

response to which Wheatstone 1838 coined the term 'stereopsis', the Greek for 'solid sight', also known as the 'plastic' effect), as well as (b) the volume of space between each object (known as the 'coulisse' effect, the French for the space between the flat-panels of stage scenery). According to this definition stereopsis is simply the perceived geometry of the scene or, as Koenderink et al. (2015b) suggest, the perception of three-dimensional space.

This definition of stereopsis contrasts with a number of authors in Philosophy (e.g. Peacocke 1983; Tye 1993) and Psychology (e.g. Hibbard 2008; Vishwanath 2010) for whom stereopsis and the perceived geometry of the scene can come apart. For instance, all four authors argue that whilst closing one eye may reduce stereopsis (i.e. lead to a reduction in our subjective impression of visual depth), it does not affect the perceived geometry of the scene (i.e. the scene itself does not appear to be any flatter).

1   TWO CONCEPTIONS OF STEREOPSIS

But however stereopsis is defined, the fundamental question is what gives rise to this subjective impression of visual depth? Over the last 150 years there has been an ongoing debate between two schools of thought:

The first school of thought regards our stereoscopic impressions as simply the product of (a) *Optical* cues, such as binocular disparity (the difference between the two retinal images), possibly with the addition of (b) *Physiological* cues, such as accommodation (the focal distance of the eyes) and vergence (the angle between the eyes), but *without* the need for the visual system to (c) attribute *contextual information* or *subjective meaning* to these cues (apart from the limited conceptual apparatus required to construct *any* 3D surface in space). According to this account the perceived geometry of the scene, which we experience as stereopsis, is simply specified by the optical information that we receive from the environment.

This conception of stereopsis is closely associated with Physiological Optics, and has been held at one time or another by Hering (1865; discussed by Turner 1994), Mach (1868, 1886; discussed by Banks 2001), Cajal (1904; cited by Bishop and Pettigrew 1986), Gibson (1950), and Julesz (1960), and still holds sway in contemporary Physiology where 'stereopsis' is often simply defined as:

> The sense of depth that is generated when the brain combines information from the left and right eyes. (Parker 2007)

...the ability of the visual system to interpret the disparity between the two [retinal] images as depth. (Livingstone 2002)

...the third spatial dimension to be extracted by comparison of the somewhat differing aspects of targets that arise when imaged from two separate vantage points. (Westheimer 2013)

By contrast, I would suggest that this stereoscopic impression of depth is not only present when we view the world with two eyes ('binocular stereopsis'), but also when we view the world with one eye ('monocular stereopsis'). This monocular impression of depth is more commonly attributed to the second conception of stereopsis. The second conception argues that in addition to (a) *Optical* cues (such as binocular disparity), and (b) *Physiological* cues (such as accommodation and vergence), the visual system either (c) also relies upon *Pictorial* cues (such as perspective and shading), whose content is geometrically unspecified until the visual system attributes meaning to them, or (d) treats *all* depth cues (the *Optical* and *Physiological* cues, as well as the *Pictorial* cues) as being unspecified until the visual system attributes meaning to them. And the meaning that the visual system attributes to these depth cues can take one of two forms: (i) *ecologically valid meaning* in the form of *prior knowledge* (typically *natural scene statistics*), or (ii) *subjective meaning* that may or may not correspond to physical reality.

This alternative conception of stereopsis is more closely associated with Cognitive Psychology, and has been held at one time or another by Ibn al-Haytham (c.1028–1038), Berkeley (1709), Helmholtz (1866), Ogle (1950), and Gregory (1966); and is closely related to both the Gestalt Psychology of the early-twentieth Century and Gombrich's (1960) 'beholder's share'. Although Cognitive Psychology has provided us with the leading articulation of this account, and indeed the leading articulation of stereopsis over the last two decades (in the form of Cue Integration, see Landy et al. 1995; Knill and Richards 1996), the argument that stereopsis is specified by the attribution of meaning is broader than Cognitive Psychology (see Albertazzi et al. 2010; Vishwanath 2005). Nonetheless, it is worth emphasising the affinity between this Psychological account of stereopsis and two of the central concerns of 'Cognitive Revolution' of the 1950–1960s:

a. Perception as Creative Construction: For Neisser (1967), the 'central problem of cognition' was the fact that visual experience is *creatively* constructed. Indeed, he coined the term 'Cognitive Psychology' to 'do justice ... to the continuously creative process by which the world of experience is constructed': 'As used here, the term "cognition" refers to all the processes by which the sensory input is transformed, reduced, elaborated, stored, recovered, and used. It is concerned with these processes even when they operate in absence of relevant stimulation, as in images and hallucinations'.

b. Centrality of the Mind: This account of perception necessarily presupposes that *the mind* (which Neisser articulated as the software of the brain, in contrast to its physiological hardware) would have a central role in determining the content of perception. Indeed, this had already been a central contention of the 'Cognitive Revolution' in the decade before, in particular the 'New Look' literature that started with Bruner and Goodman (1947). As Miller (2003) explained of the 'Cognitive Revolution': 'We were still reluctant to use such terms as 'mentalism' to describe what was needed, so we talked about cognition instead'.

2   A RECENT HISTORY OF STEREOPSIS

But whilst Psychology was undergoing a revolution to give the mind a central role in determining perceptual content, stereopsis was about to undergo a transformation in the opposite direction. For instance, Bishop and Pettigrew (1986) draw a distinction between the pre-1960 literature on stereopsis:

Stereopsis Before 1960: Mentalism Prevails

And the post-1960 literature on stereopsis:

The Retreat From Mentalism Begins In The 1960s

As Bishop and Pettigrew explain, the pre-1960 literature on stereopsis was an unbroken linage that stretched for a century from Helmholtz (1866) to Ogle (1959). Speaking of the period before 1960, Bishop and Pettigrew observe:

Before that time it was generally believed that binocular depth perception was based on high-level quasi-cognitive events that took place somewhere in the no-man's land between brain and mind.

For instance Ogle, the leading authority on binocular stereopsis at the time, wrote in his introduction to *Researches in Binocular Vision* (1950) that depth perception was a synthesis of (a) the retinal stimulation, (b) the physiology of the retina and the neurological processes by which this retinal stimulation was communicated to the brain, (c) the 'psychic modifications and amplifications' of these 'neurologic "images"' by past visual, auditory, and tactile experiences, and (d) the modifying effects of attention and the motivations of the individual.

Indeed, as late as the 1950s some Gestalt Psychologists still argued that stereopsis was a *purely* Psychological phenomenon (Ogle 1954 cites Tausch 1953, and Ogle 1959 cites Wilde 1950). Although Ogle rejected this extreme position (Ogle 1954), he nonetheless insisted that the *meaning* attributed to pictorial cues by the visual system could modify or even dominate the stereoscopic impression from binocular disparity. For instance, Ogle (1959) distinguishes between (a) stereopsis from binocular disparity, which he regards as automatic, and therefore *meaningless*, and (b) 'empirical clues' such as perspective, which have been made *meaningful* by experience. And he goes on to conclude:

> ...it is to be expected that in those surroundings that have been artificially produced to provide a conflict between stereoscopic stimuli and empirical factors, the meaningless stimuli may be suppressed by the meaningful, that is, by the perceptions from the empirical motives for depth.

This passage was heavily influenced by Ogle's mentor at Dartmouth, Adelbert Ames Jr. (1951, 1955). Indeed, Ogle advances Ames' Window (where a trapezoid window frame constructed to look like a rectangular frame seen in oblique perspective appears to change direction as it is rotated) as just such an instance of *meaningful* empirical cues (in this case perspective) dominating *meaningless* binocular disparity.

But the impetus for stereopsis' anti-Cognitive Revolution in the 1960s was not Ogle's (1959) suggestion that binocular disparity could be modified by pictorial cues, but instead his insistence that depth could not be extracted from binocular disparity until figure-ground meaning had first been attributed to both of the retinal images:

> We must stress the importance of contours, those lines of demarca-
> tion between the 'figure' and the 'background.' In every case stereo-
> scopic depth depends on the disparity between the images of identifiable
> contours.

It was in response to this claim that Julesz (1960) created the Random-
Dot Stereogram (see also Aschenbrenner 1954). Julesz had been a radar
engineer in the Hungarian military where the practice had been to use
stereo-images (two images taken from different perspectives) to break
camouflage in aerial reconnaissance: the camouflaged object was indiscriminable until the images were fused stereoscopically, at which point
the object would jump out in vivid depth. Julesz therefore hypothesised
that stereopsis must *precede* contour recognition, and he invented the
Random-Dot Stereogram as a form of 'ideal camouflage' to prove this
very point: comprised of two images of apparently randomly distributed dots, the contours of a hidden object are encoded in the differences
between the images (rather than the individual images themselves), and
yet the hidden object emerges in vivid depth when they are fused (Fig. 1).

The Random-Dot Stereogram not only revolutionised the understanding of stereopsis in Psychology, it also had a significant impact upon
Neurophysiology by inspiring the search for 'disparity selective neurons'
that could track these differences between the two images (see Pettigrew
1965). Indeed, as Cumming and Parker (1999) observe, the discovery
of disparity selective neurons by Barlow et al. (1967) and Nikara et al.
(1968) would further conflate stereoscopic depth perception and binocular disparity.

But the problem with equating stereopsis and binocular disparity in
this way is the implication that monocular vision lacks this subjective
impression of depth. For instance, in his Ferrier Lecture on Stereopsis,
Westheimer (1994) insisted that 'real stereo sensation is absent with
monocular viewing', whilst Parker (2016) appears to suggest that 'a
direct sense of depth' only emerges with binocular vision. Although this
position continues to have adherents in Physiology, and is attractive to
others on purely experiential grounds (see Sacks 2006; Barry 2009; and
Sacks 2010), by the mid-1990s it had come to be rejected by Cognitive
Psychology. As Landy et al. (1995) observe, in what is arguably the most
influential paper of this period, if you close one eye the world does not
suddenly become flat. The implication being that stereopsis is not just a

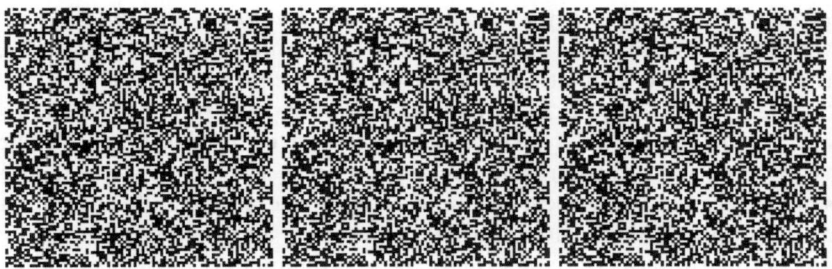

Fig. 1 Random-Dot Stereogram proposed by Julesz (1960). A square appears in stereoscopic depth when you cross-fuse the left and central image (by focusing on a point in front of the image) or parallel-fuse the central and right image (by focusing on a point behind the image) in spite of the fact that there are no contours demarcating the square.

product of binocular disparity but also monocular cues to depth such as shading, perspective, and occlusion. Over the last couple of decades this observation has been explored in two distinct ways:

1. Cue Integration: The first strand in the literature marks a return to Ogle's (1959) observation that monocular depth cues can modify the stereoscopic impression from binocular disparity. As Landy et al. (1995) and Knill and Richards (1996) documented in the mid-1990s, when faced with multiple sources of depth information the visual system appears to combine this information into a single coherent percept either by integrating the information linearly (if the sources of information are only mildly in conflict) or by down-weighting or excluding apparently aberrant sources of information (when the conflicts are large).

As we shall see in Chap. 2, this process of Cue Integration has been systematically tested and confirmed in the literature using cue-conflict stimuli. I do not seek to challenge most of these results. Instead, what I do seek to challenge is the *interpretation* that has been given to these results. To explain why, we should recognise that there is at least one part of the Cognitive Revolution that, some 60 years on, still feels incomplete: as Miller (2003) observed, the Cognitive Revolution was a reaction against the excesses of Behaviourism, according to which *perception* was equated with *discrimination*; *memory* was equated with *learning*; and *intelligence* was equated with *what intelligence tests test.* And yet, even to this day, the tendency to conflate *perception* with *discrimination* still persists. So I argue in Chap. 2 that many cue-conflict experiments appear to reflect their subjects' (mistaken) *judgements* or *evaluations* of their visual experience, rather than the depth that is actually perceived. The same, I argue, also appears to hold true of cue-conflict illusions such as the hollow-face illusion and Reverspectives, and I suggest that they are better thought of as *delusions* (false judgements) rather than *illusions* (false percepts).

2. Monocular Stereopsis from a Static 2D Image: The second implication of depth from pictorial cues is that we should be able to experience monocular stereopsis by viewing a single static 2D image with one eye closed. This is the concern of a second strand in the literature that started in the mid-1990s with the exploration of so-called paradoxical monocular stereoscopy by Koenderink et al. (1994) (see also Enright 1989, and Eby and Braunstein 1995); and has most recently been explored by Koenderink et al. (2013), Vishwanath and Hibbard (2013), Volcic et al. (2014), Vishwanath (2016), and Wijntjes et al. (2016). This literature also revived a long forgotten tradition of monocular stereoscopy by some of the finest minds in early-to-mid twentieth century vision science: von Rohr (1903), Claparède (1904), Holt (1904), Münsterberg (1904), Ames (1925a, b), Carr (1935), Schlosberg (1941), Gibson (1947), and Gabor (1960).

It is tempting to suggest that this early literature was a casualty of Julesz (1960), and the subsequent conflation of stereopsis with binocular disparity. But in truth monocular stereopsis had already been rejected by Ogle (1959) on purely experiential grounds. Ogle suggested that Wheatstone's (1838) stereoscopic line drawings demonstrated the 'fundamental' difference between binocular depth *perception* on the one hand, and the mere *conception* of depth available from monocular viewing: he argued that there was absolutely no impression of stereoscopic depth from these simple line drawings when they were viewed monocularly, and yet when they were viewed in a stereoscope such images produced a vivid impression of depth. This is true, but Ogle's mistake was to generalise from this example to all instances of monocular depth: he used this example to conclude that stereoscopic depth perception was the 'single outstanding function of vision with the two eyes', that was 'not even suggested by vision with one eye alone.'

According to the contemporary literature, Ogle's mistake was to rely on simple line drawings as being representative of all 2D images: it would argue that once depth cues such as perspective and shading are added to the image, there is ample evidence that (a) monocularly viewed images produce a depth percept, and also (b) that synoptic viewing (viewing two identical 2D images in a stereoscope) can significantly accentuate this impression of depth: see Koenderink et al. (1994) and Wijntjes et al. (2016). On the one hand, in Chap. 3 I question whether the evidence in favour of monocular depth perception from 2D images really goes to our *perception* rather than our *cognition* of depth. But on the other hand, I also resist Ogle's suggestion that if we fail to perceive depth in a monocularly viewed 2D image, this must also imply that our monocular vision of the 3D world also lacks depth.

3. Monocular Stereopsis in a 3D World: But how might we explain the monocular perception of depth in the 3D world if it is absent in a 2D image? Well, objects distributed throughout space in a 3D world will be subject to various different degrees of defocus blur. Traditionally when defocus blur has been treated as a depth cue, it has been regarded as just another pictorial cue alongside perspective and shading. But for this pictorial account of defocus blur to work it has to penetrate our subjective visual experience. Consequently, since defocus blur is typically apparent less than 4% of the time, it is commonly assumed that defocus blur must be a depth cue with only limited application: see Sprague et al. (2016). By contrast, in Chap. 4 I argue that if my contention in Chap. 3 is correct, and we do not perceive depth from perspective or shading, then we need another explanation for why we appear to be able to see depth when we look at the 3D world with one eye. The only solution, I suggest, is that just as the visual system can rely on sub-threshold defocus blur in order to guide accommodation, it can also rely on sub-threshold defocus blur in order to determine, in a very rough sense, the depth relations in the scene.

Now whilst sub-threshold defocus blur might account for the perceived geometry of a monocularly viewed scene, what about its scale?

A common assumption is that we can scale a monocular scene using accommodation (the tension in the ciliary muscles that control the intraocular lens indicating the distance of the focal plane). But as I argue in Chap. 4, there are both theoretical and empirical considerations that militate against this hypothesis. Instead, I conclude that we should be open to the idea that monocular vision does not convey scale, and that scale is only something that we *cognitively impute* to the scene. Indeed, Chap. 4 raises the prospect that this might hold true for binocular vision as well.

4. Extracting Depth from Binocular Disparity: But what about the claim that started the anti-Cognitive Revolution in the first place, namely Ogle's insistence that figure-ground separation was a prerequisite for extracting depth from binocular disparity? Well, so far as this question is concerned, if the 1960s had marked a 'retreat from mentalism', then the neo-Gestalt Revolution of the 1980s (see Ramachandran 2006) marked a return: although Julesz (1960) was right that figure-ground separation was not a *prerequisite* for extracting depth from disparity, Ramachandran and Cavanagh (1985) demonstrate that the subjective contours of Kanizsa figures (see Fig. 7) appear to influence to the structure of the depth that is derived from binocular disparity (see also Ramachandran 1986; Nakayama et al. 1989; and Mather 1989, although Mather leaves it open whether subjective contours really influence the extraction of depth from disparity, or are themselves merely a consequence of it).

Furthermore, Zhou et al. (2000) argue that V2 (the secondary visual cortex) not only identifies contours but also differentiates between figure and ground, whilst Qiu and von der Heydt (2005) go one step further by suggesting that V2 achieves this differentiation between figure and ground by employing disparity and Gestalt rules alongside one another (see also Nakayama 2005; Ramachandran and Rogers-Ramachandran 2009; and von der Heydt 2015).

The suggestion that *meaning* can be attributed to monocularly viewed images by the visual system, and that this meaning can be used to disambiguate the signal from binocular disparity, is an intriguing one; but if my analysis of monocular vision in the previous subsection is correct, its importance is liable to be overstated:

First, we would have to ensure that Mather's (1989) alternative explanation for this phenomenon had been entirely ruled out. It is plausible that the subjective contours are experienced in stereoscopic depth not because (a) figure and ground have been interpreted in the monocular image before the extraction of depth from disparity, but because (b) this is the most parsimonious solution as to how the sparse stereo elements fit together coherently in a 3D scene.

Second, even if it turns out that figure-ground separation is relied upon to disambiguate stereograms with sparse disparity information, such disambiguation is *2D-plus* rather than *3D*: it involves (a) the 2D *segmentation* of the image, followed by (b) the *ordering* of its layers, but this is still far removed from (c) the attribution of *depth* to these layers. In this sense, figure-ground cues are more like the recognition of words on a page than the attribution of depth: words are recognised as being *on* the page, even though there is *no depth* between the words and the page.

Third, to the extent that the literature on disambiguating binocular disparity claims anything more, and moves from (a) using monocular cues to *disambiguate* binocular disparity to (b) placing monocular cues *in conflict* with binocular disparity (see Qiu and von der Heydt 2005), then it strays into the cue-conflict literature which is fully explored in Chap. 2 (this is the reading of Qiu and von der Heydt 2005 advanced by Burge et al. 2010). Indeed, we might use the cue-conflict literature to try to better understand which side of the perception–cognition divide V2 lies (when conjoined with the results in Qiu and von der Heydt 2005, my position in Chap. 2 logically entails that V2 is engaged in cognition rather than perception).

Finally, even if the visual system relies upon subjective contours to extract depth when faced with (a) two flat 2D images, and (b) sparse disparity information by which reconcile these two images, I am sceptical that Ramachandran and Cavanagh's (1985) findings have any general application outside this context. Under my account, the real world already provides the visual system with an optical cue to figure-ground relations: defocus blur. So stereoscopic viewing of 2D images with sparse disparity information begins to look like an artificial, contrived, and arguably misleading, basis upon which to understand the relationship between stereopsis and pictorial processing.

Indeed, as Mach recognised over 140 years ago, if you already have a monocular conception of stereopsis, then binocular stereopsis begins to resemble a secondary process that merely accentuates the prior monocular processing (see Banks 2001). Similarly, Koenderink et al. (2015b) describe monocular stereopsis as 'stereopsis proper' and suggest that binocular stereopsis is at least partly, but probably largely, to be explained in monocular terms. Such a conclusion would also make sense from an evolutionary perspective: monocular stereopsis must have emerged in herbivores before binocular stereopsis emerged in predators. Consequently, we would expect the binocular depth processing that emerged to be parasitic upon the monocular depth processing that already existed.

3   VISUAL COGNITION

But even if I am right and pictorial cues do not contribute to our perception of depth, the pictorial cues in cue-conflict stimuli (Chap. 2) and 2D images (Chap. 3) clearly contribute to *something*: if it is not our *perception* of depth, then what is it? I would argue that they contribute to an *automatic* (i.e. not consciously or deliberately made, and often involuntary) *post*-perceptual *evaluation* (or *judgement*) of the scene. Under this account pictorial cues are not *perceptual* cues but *cognitive* cues. But, and this is the important point, they are cues to a relatively self-contained module of *cognition*, divorced from *conscious deliberation*.

In this sense, there is an affinity between my position and Cavanagh's (2011) account of *visual cognition* as an *unconscious* (we are unaware of it at work), *automatic* (we do not have to do anything), and *involuntary* (we often cannot overrule it) process that attributes *meaning* to sensory data *before* conscious deliberation. But the key difference is that for Cavanagh, visual cognition operates as at the level of perception (Fig. 2).

Cavanagh (2011) documents the 'extraordinarily sophisticated' perceptual inferences of visual cognition that are distinct from conscious deliberation. The classic example is the Müller-Lyer illusion (see Chap. 2, Fig. 12): the illusion still persists even though we *know* that the lines in the Müller-Lyer illusion are the same length. But whilst Pylyshyn (1999) interprets the persistence of the Müller-Lyer illusion as evidence of *cognitive impenetrability* (i.e. of perception being immune from cognitive influence), Cavanagh insists that it actually evidence of *cognitive independence*:

> Pylyshyn calls this cognitive impenetrability but we might see it as cognitive independence: having an independent, intelligent agent—vision—with its own inferential mechanisms.

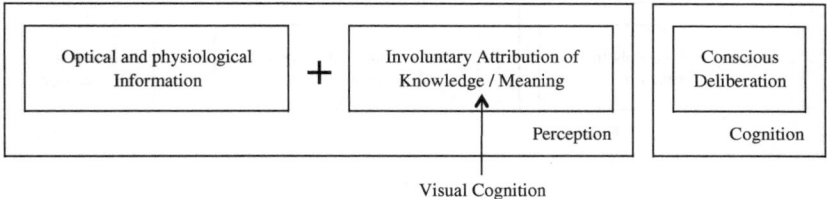

Fig. 2 Visual cognition according to Cavanagh (2011)

For Cavanagh, the inferential mechanisms of visual cognition play an essential role in determining the content of perception. Whether they are the top-down or high-level inferences of Bayesian Cue Integration, or merely the mid-level inferences associated with Gestalt Psychology, the point is the same: the retinal information is insufficient to specify the percept, so inferential mechanisms are relied upon to determine which percept out of the many possible percepts we in fact see. This is not to constrain the form these inferences must take: as Cavanagh observes, they may be based on likelihood, bias, or even a whim. But the important point is that whatever form these inferences take, the visual system uses them to reject the many possible alternatives that were just as consistent with the raw sensory data as the eventual percept.

Nor is Cavanagh's account in tension with Firestone and Scholl's (2016a, b) recent work on cognitive impenetrability. Whilst Firestone & Scholl are refreshingly robust about the need to distinguish between *perception* and *cognition*, by *cognition* they mean the thoughts, desires, and emotions of the New Look literature (see Bruner and Goodman 1947). They are motivated by the *'revolutionary* possibility' that what we see is directly influenced by what we think, want, and feel. By contrast, Firestone & Scholl explicitly exclude the unconscious inferences that Cavanagh has in mind as being in any way controversial or suggestive of cognitive penetration, claiming that a good litmus test is whether such inferential processes continue to operate reflexively in spite of our own conscious deliberations (as is the case with the vast array of visual illusions, from the Müller-Lyer illusion to the hollow-face illusion and Reverspectives).

Now whilst I agree with Cavanagh (2011) that there appears to be a relatively self-contained module of visual cognition, I would argue (at least in so far as depth perception is concerned) that this module ought

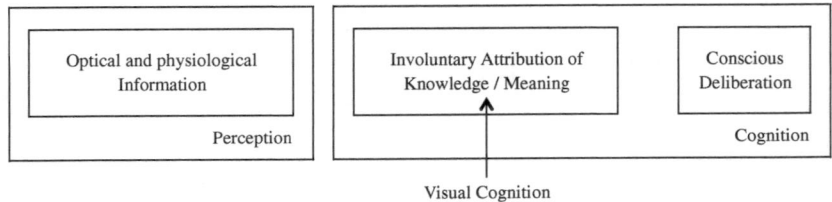

Fig. 3   Visual cognition according to my alternative account

to be regarded as *post-perceptual*, since it does not appear to affect the actually perceived geometry of the scene, but only our *judgement* or *evaluation* of it (Fig. 3).

A useful analogy I explore in Chap. 3 is with *reading*: being able to understand the meaning of a word doesn't change the perceptual appearance of the text, but nor does it rely upon conscious deliberation. Instead, it appears to be both a *post-perceptual* and an *automatic* and *involuntary* process of attributing meaning to what we see. Similarly, I would argue that something 'looking flat', 'looking round', 'looking square', or 'looking symmetrical' is not really a *perceptual* claim, but a *post-perceptual* attribution of depth or shape meaning: a *judgement* or *evaluation* about what we see. And I would argue that pictorial cues *bias* our evaluation of depth, rather than informing our perception of depth. On the other hand I agree with Cavanagh that visual cognition must be *pre-deliberative*, since it biases our evaluation of depth in a way that is apparently *not* open to rational revision (in this sense *visual cognition* is not only *automatic*, but also *involuntary*). Indeed, often the only way to counteract these biases is to introduce a visual comparator (see Chap. 2); in a sense, to change the cognitive task from an *evaluation* to a simple *comparison*.

This debate is not only important for depth perception, but also the wider question of the role of cognition in vision. After all, depth from pictorial cues represents the thin edge of a very significant cognitive wedge for Cavanagh. And, as more and more complex phenomena (such as *causation* and *intentionality*) are attributed to vision, the more intelligent Cavanagh insists the visual system must be:

> ...the unconscious inferences of the visual system may include models of goals of others as well as some version of the rules of physics. If a 'Theory

of Mind' could be shown to be independently resident in the visual system, it would be a sign that our visual systems, on their own, rank with the most advanced species in cognitive evolution.

By contrast, one of the virtues of my account is that we do not have to posit the existence of 'an independent, intelligent agent—vision' to explain these increasingly complex phenomena; instead, we simply recognise that post-perceptual human cognition may be broken into relatively independent modules.

4   FOUR CONCEPTIONS OF MEANING

In this final section, I outline four of the leading accounts of stereopsis: Pictorial Cues, Cue Integration, Gestalt Psychology, and Intentionality, and explore the kinds of meaning that each account suggests must be attributed to the raw sensory data before content can be extracted from it and/or attributed to it:

1. Pictorial Cues: I will explore perspective and shading as a means by which to understand the extraction of depth from pictorial cues more generally:

(a) Perspective: In 1903, Moritz von Rohr developed 'The Verant, a New Instrument for Viewing Photographs from the Correct Standpoint' for Carl Zeiss based upon the work of Allvar Gullstrand (see von Rohr 1903). This monocular lens ensured that observers could view a 2D image from its centre of projection, and for von Rohr this (in addition to setting accommodation at infinity) explained the impression of monocular depth that subjects reported: by placing their eye at the centre of projection, the subject experienced the very same perspective cues that they would have experienced had their eye been placed at the entrance pupil of the camera.

This claim was explored by Holt (1904) and Schlosberg (1941). As a disciple of Holt, and a close associate of Schlosberg's, Gibson (1947) could not ignore the implications of this observation. In his work for the US military during WWII, he agreed that if a single static 2D image was viewed monocularly, whilst eradicating any cues to flatness, the observer was liable to experience a monocular impression of visual depth equivalent to binocular stereoscopic viewing. Indeed, Gibson (1947) drew the conclusion that if binocular disparity appeared to contribute little to our impression of depth from 2D stereoscopic images, then it must also contribute little to our impression of depth from the 3D world, and this led Gibson (1950) to embrace an account of depth perception according to which binocular disparity played a largely insignificant role.

But we still need to explain why a static 2D image viewed from its centre of projection should induce an impression of depth? For von Rohr the answer was clear: viewed in this way, the observer experiences the very same perspective cues they would experience had they been present in the real world scene. But as Gibson observed, this explanation only poses a further problem: namely, why should perspective cues *from a real world scene* give rise to a monocular impression of depth in the first place? Gibson toyed with this question for much of his 50-year career, although the emphasis appears to shift away from monocular stereopsis towards pictorial depth: for instance, Schlosberg (1941) is cited in Gibson (1966) but not Gibson (1971) or Gibson (1979). One gets the impression that Gibson never fully resolved this question to his satisfaction. As he recounted just before his death (in Gibson 1979), he repeatedly revised his theory of pictorial cues, leaving a catalogue of abandoned accounts: Gibson (1954, 1960), and Chap. 11 of Gibson (1966). The intractable problem for Gibson (1979) was that perspective is *indeterminate*: it might specify *some* invariant features the scene must have, but it is neutral as between the various competing arrangements that satisfy these features. Indeed, this realisation led Gibson (1979) to ultimately reject monocular stereopsis from a 2D image, an insight that had previously meant so much to him:

> The purveyors of this doctrine disregard certain facts. The deception is possible only for a single eye at a fixed point of observation with a constricted field of view... This is not genuine vision, not as conceived in this book.

And yet for contemporary neo-Gibsonians, Gibson's most difficult case turns out to be their easiest. Consider Rogers and Gyani's (2010) discussion of Patrick Hughes' 'Reverspectives', *protruding* physical forms that are painted as if they are *receding* in perspective (in this instance, the canals of Venice) (Fig. 4).

Fig. 4 Patrick Hughes in his studio. © Patrick Hughes. For more information please see: http://www.patrickhughes.co.uk/

Rogers and Gyani (2010) suggest that when stationary observers view this artwork monocularly, they perceive it as a scene *receding* in depth rather than its actual physical form (i.e. as an object *protruding* in depth). For Rogers and Gyani, the reason for this depth inversion is 'obvious': 'What we see is consistent with the information provided by the perspective gradients'. But the question is not whether the illusory percept is *consistent* with perspective. That is a given. Instead, the question is why *this* percept is chosen out of the innumerable consistent possible interpretations? This was Gibson's question. And for Rogers and Gyani, the answer is that the illusory percept is not just *consistent* with perspective, but *specified* by it:

> ...we should not be surprised that we see 'reversed' depth when these delightful artworks are viewed monocularly *because this is what the perspective information is telling us...* (emphasis added)

By contrast, I would argue that there is no such thing as *perspective information*, only *optical information* to which *perspective meaning* has been attributed.

Indeed, *perspective meaning* is something that has to be learnt. This is demonstrated by the fact that perspective images mean nothing to those with newly restored sight if they have been blind all their lives. For instance, when Sidney Bradford had his sight restored (see Gregory and Wallace 1963) he was immediately able to understand capital letters and clock-faces (as they had been taught to him via touch) but not, as Gregory (2004) explains, pictures: pictures looked flat and meaningless to him, in spite of the fact that he could judge the size and distance of objects that were already familiar from touch (e.g. chairs scattered around the ward). Furthermore, the process of learning to attribute meaning to perspective is gradual: even after six months Mike May, another formerly blind patient, was unable to identify wireframe drawings of cubes in any orientation, describing them as 'a square with lines' (see Fine et al. 2003).

So whilst Rogers and Gyani (2010) may dismiss the experience of a static monocular observer of a Reverspective as uninformative (at one point suggesting that it 'cannot tell us anything about the visual system'), I would argue that it would, in fact, tell us something very significant: namely that (if Rogers and Gyani are correct) the visual system utilises a learnt form of *meaning*, perspective meaning, to determine the content of stereopsis. You might object that calling this *meaning* puts the point too finely. After all, Rogers and Gyani are keen to emphasise the low-level nature of perspective: they demonstrate that a simple wireframe Reverspective can be just as effective as a fully rendered scene, and suggest that converging line junctions are sufficient to induce a depth percept all by themselves.

Similarly Cavanagh (2011) excludes low-level processes from visual cognition. Indeed, there is a similarity between the way subjective completion can lead to perverse results (for instance, the conjoining of the front of one animal with the back of another into an impossibly long form), and the way in which the visual system is liable to come to an automatic interpretation of perspective cues, even if it is obviously wrong (for instance, in the context of an Ames Rooms). In the context of subjective completion, Cavanagh asks: 'Given this very lawful behaviour, we might ask if there is anything really inferential here'. The same, Rogers

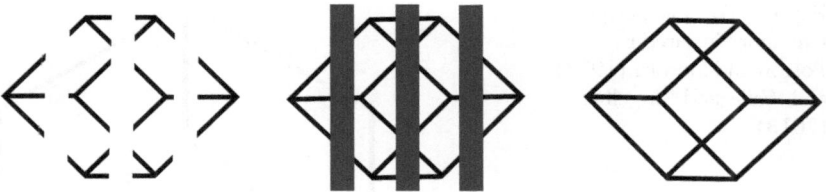

Fig. 5 Isolated line junctions inspired by Nakayama (1999)

and Gyani would argue, could be asked in the context of extracting depth from perspective.

But I think it would be a mistake to draw a distinction between *lawful* processing on the one hand and *cognition* on the other. Cavanagh's account is liable to run three distinct concerns together: (a) *inference* (incorporating some notion of *problem-solving*), (b) *complexity* (incorporating some notion of *intelligence*), and (c) *choice* (incorporating some notion of *agency*). By contrast, I think *agency* is unhelpful in this context: logic, mathematics, and linguistics are all forms of rule-based reasoning that clearly ought to qualify as cognition if the visual system is engaged in them. Nor can we dismiss the rule-based extraction of depth from perspective as *just processing*, since everything in the brain is ultimately 'just' rule-based processing. Indeed, this is reflected in Cavanagh's own description of inferences:

> Note that an inference is not a guess. It is a rule-based extension from partial data to the most appropriate solution.

So the choice is between *simple* rule-based processing and *complex* rule-based processing; Rogers and Gyani (2010) may be an instance of the former, and Cavanagh (2011) primarily concerned with the latter, but the *complexity* of the meaning being attributed makes no difference to my account: rudimentary meaning is still meaning.

But my second response to Rogers and Gyani is to question just how rudimentary the extraction of depth from perspective really is? Whilst Rogers and Gyani suggest that the visual system exploits line junctions, Nakayama (1999) demonstrates that line junctions all by themselves are not necessarily that informative (Fig. 5).

Fig. 6 Tristable perspective figure inspired by
Poston and Stewart (1978)
and Wallis and Ringelhan
(2013)

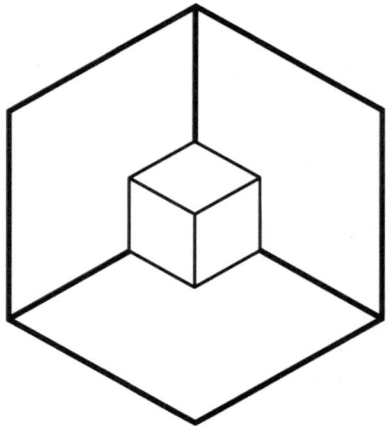

First, these junctions can easily be given a 2D interpretation. Second, even if they are given a 3D interpretation, it is far from obvious that they represent three angles of equal size. Instead, this interpretation only appears to emerge once the individual junctions are themselves seen as part of a coherent whole: it is as if the eight junctions become eight simultaneous equations, to which 90° is the only rational solution. But if this is the case, and the *whole* specifies the *parts*, then this is far from a *low-level* process. Indeed, we reach the same conclusion by consider- ing multi-stable cubic volumes whose components are liable to be inter- preted as a coherent whole (either as a small cube against a background or as large cube with a small chunk taken out of it) even though a small perspective cube in front of a large perspective cube is just as permissible an interpretation (Fig. 6).

But the deeper concern is that by focusing on a 'carpentered world' of parallel lines and right-angles (such as cubic volumes, Ames Rooms, and Reverspectives), we risk massively underestimating the complexity of the processes that extract depth from perspective. It is easy to forget that the visual system did not evolve in response to a 'carpentered world', and that the forms it did evolve in response to were positively irregular by comparison. Consequently, the visual system's response to perspective cues is likely to be much more nuanced than its automatic interpreta- tion of cubic volumes would suggest. Indeed, we do not need to appeal to the positively irregular forms of human evolution to illustrate this

point: as Knill (2007) demonstrates, even extracting perspective information from a regular shape like an ellipse depends heavily on prior knowledge and/or assumptions about the most likely interpretation of a given
scene. Landy et al. (2011) explain the kind of complex scene statistics
that would have to be employed by subjects:

> The generative model for the aspect ratio of an ellipse in the image
> depends on both the 3D slant of a surface and the aspect ratio of the
> ellipse in the world. The aspect ratio of the ellipse in the world is a hidden
> variable and must be integrated out to derive the likelihood of slant. The
> prior distribution on ellipse aspect ratios plays a critical role here. The true
> prior is a mixture of distributions, each corresponding to different catego-
> ries of shapes in the world.

Furthermore, according to Knill (2007) the visual system doesn't just
engage in natural scene statistics, it also engages in real-time *perceptual
learning*: The subjects in Knill (2007) initially assumed that any ellipse
in the visual field must be a slanted circle. But as the experiment pro-
gressed they encountered a number of patently non-circular ellipses,
and so learnt that the circularity of the ellipses could not be assumed.
Consequently, when the slightly non-circular ellipses that had previously been judged as circular earlier in the experiment were shown again,
the subjects now correctly identified them as non-circular.

b. Shading: So extracting depth from perspective proves to be anything but a low-level process, and the same appears to be true for
shading:

First, shading is a change in the *luminance* of a surface, but our interpretation of surface luminance is a complex phenomenon, that is only
partly determined by the amount of light that is reflected from the object
to the retina: see Gilchrist (2006).

Second, extracting depth information from *changes* in luminance
requires a mechanism that can take those changes into account. But
Tyler (2006) suggests that such a mechanism would have to be surprisingly complex; certainly well beyond the range of early visual image fil-
ters which Morgan and Watt (1982) have estimated only extends to 2–3
arc min (1/30th to 1/20th of a degree). So if shape is interpolated on
the basis of changes in luminance over the surface of the object, a mid-
level or higher-level process must be responsible.

Third, accurately extracting depth information from shading requires prior knowledge about the direction of illumination (see Pentland 1982). For Wagemans et al. (2010), this is evidence that 'the shading cue is inherently ambiguous', leading them to give up on inverse optics in Koenderink et al. (2015a) and instead treat shading merely as an instance of 'relief articulation', much like contour-lines drawn on a map to convey relief. The only alternative is to appeal to a default assumption about the illumination in the scene. Three candidates have been advanced: The first is to posit a single strong *overhead* light-source (i.e. the sun): see Ramachandran (1988). The second is to suggest that light, having been reflected between the atmosphere and the ground multiple times, is *diffuse*: see Gibson (1979) and Chen and Tyler (2015). The third is to adopt an *ecological* perspective, according to which both are permissible: overhead light on a sunny day, and diffuse light on a cloudy day; but this entails an even more sophisticated process of extracting depth from shading given that overhead and diffuse light cast such very different shadows, see Langer and Bülthoff (2000).

Fourth, once we have finally settled on an appropriate assumption, we still have to use it to extract the relevant depth information from shading, and this promises to be another complex undertaking: Tyler (1998) and Chen and Tyler (2015) have argued that under the diffuse light assumption the visual system can adopt a quick and easy 'dark is deep' rule of thumb, but Langer and Bülthoff (2000) and Todd et al. (2015) have demonstrated that even under diffuse light dark does not necessarily mean deep, and so have questioned the ecological validity of this approach.

2. Cue Integration: As these discussions illustrate, the process of extracting depth information from a single depth cue such as perspective or shading implies a significant degree of complexity. Consequently, individual depth cues are liable to provide us with only partial, noisy, or contradictory depth information. But if this is the case then a second stage of cognitive processing is required in order to integrate and reconcile these various contradictory sources of depth information into a single coherent percept. And prior knowledge is thought to play a central role in this integration process.

In the contemporary literature this reliance on prior knowledge is typically articulated in Bayesian terms, and there is no doubt that the Bayesian literature of the last couple decades has brought greater statistical sophistication to bear on this question. Nonetheless, as Trommershäuser et al. (2011) observe, the fundamental principle that underpins Cue Integration was already apparent in Helmholtz's (1866) unconscious inferences, and even in the work of al-Haytham (c.1028–38). Similarly, Seydell et al. (2011) suggest that we might regard Cue Integration as the *veridical* counterpart to Gregory's (1970) hollow-face *illusion*: whilst the visual system's reliance upon prior knowledge may give rise to illusions in certain artificially contrived contexts (e.g. the misinterpretation of a hollow mask), ordinarily a reliance on prior knowledge only improves the visual system's ability to estimate the true state of the world.

For Cue Integration this reliance on prior knowledge is a *prerequisite* for perception. This is sometimes overlooked in the literature, where there can be a tendency to pit 'top-down' prior knowledge against 'bottom-up' sensory data. For instance, Nguyen et al. (2016) suggest that what we see depends upon two types of influences that can be in competition: (a) 'bottom-up' cues such as edge orientation, the direction and speed of motion, luminance and chromatic contrast, and binocular disparity, and (b) 'top-down' influences such as endogenous attention, expectations, and stored visual knowledge, of which they advance Bayesian Cue Integration as an example. But to suggest that 'top-down' processing is either in conflict with, or merely influences, the 'bottom-up' sensory data is to underestimate the importance of 'top-down' processing for Cue Integration accounts: according to Cue Integration 'top-down' processing is the *only* way of attributing depth meaning to sensory data, without which the sensory data would simply have no content. So it is not as if 'top-down' processing merely *influences* or *competes with* the 'bottom-up' sensory data, or that if the 'top-down' processing were absent 'bottom-up' sensory data would be free to determine the percept; instead, 'top-down' processing constitutes perception under a Cue Integration account.

Finally, although Trommershäuser et al. (2011) observe that 'Bayesian statistics is emerging as a common language in which cue-combination problems can be expressed', this is not the only articulation of Cue Integration in the literature. Indeed, since the late 1990s Domini, Caudek, and colleagues have emphasised the *non-veridical* and often *mutually inconsistent* nature of Cue Integration (see Domini and

Caudek 2011). Especially important for Domini and Caudek (2011), as well as for Scarfe and Hibbard (2011), is the possibility that individual cues might be *biased*. Domini and Caudek argue that if it can be demonstrated that bias really is pervasive in the visual system, then this should have a transformative effect on how we ought to conceive of vision: Is the goal of vision to recover a veridical depth map of the scene? Or is it, as Domini and Caudek suggest, merely concerned with ensuring that we can effectively interact with the environment?

Indeed, this concern with successful interaction, rather than recovering a metric depth map, reflects a recent trend in cognitive science which Engel et al. (2013) term *the pragmatic turn*. As Engel et al. (2016) explain:

> Cognitive science is witnessing a pragmatic turn away from the traditional representation-centered framework of cognition towards one that focuses on understanding cognition as being 'enactive.' The enactive view holds that cognition does not produce models of the world but rather subserves action...

But even those who continue to articulate Cue Integration in *representational* terms are liable to (a) question the wisdom of associating Cue Integration with *optimality*: see Rosas and Wichmann (2011), or (b) suggest a *less formulaic* approach, according to which Cue Integration is closer to testing a hypothesis (see Gregory 1980 and Tyler 2004) or playing 20 questions with nature (see Kosslyn 2006 and Cavanagh 2011).

3. Gestalt Psychology: 'Gestalt' is German for 'pattern' or 'shape', although 'configuration' is closer to what was intended (see Rock and Palmer 1990), with the central argument of Wertheimer's (1924) principle of 'holism' being that we directly perceive *configurations* or *integrated wholes* whose properties are greater than the sum of their parts. The classic illustration of this is the Kanizsa Triangle (Fig. 7): The unavoidable impression is of a white triangle occluding three black circles and a wireframe triangle. But no white triangle is specified by the stimulus. Nor are the black circles or the wireframe triangle. And there is no sense in which an inverse optical account that failed to specify these circles and triangles would be incomplete. So this is taken as evidence that something more than inverse optics must be going on. And the unresolved question of the last couple of decades is what this something

Fig. 7   Kanizsa triangle
inspired by Kanizsa
(1955)

more is, and how exactly it relates to Cue Integration? Specifically, are Gestalt phenomena such as the Kanizsa Triangle (a) *an alternative to*, (b) *supplementary to*, or (c) simply *just an application of* Cue Integration?

As Wagemans et al. (2012a) observe, most textbooks will contain a chapter on Gestalt phenomena but leave their relationship with the rest of the literature ambiguous. But in another sense it is no surprise that this tension between these Gestalt phenomena and Cue Integration has not been resolved because we are still unsure as to what exactly is driving the Gestalt phenomena in the first place: is it *likelihood* or is it *simplicity*? As Wagemans et al. (2012a, b) ask, do we see the white triangle in Fig. 7 because it is the *most likely* interpretation of the stimulus, or merely because it is the *most straightforward* one?

If it is the former, then Gestalt principles are subsumed under Cue Integration. Certainly Wagemans et al. (2012a) would not shy away from this conclusion, suggesting that groupings could be based on probabilistic models derived from natural scene statistics. Indeed, even those who embrace the alternative principle of *Prägnanz* (or *simplicity*) are often inspired to do so by Structural Information Theory on the basis that in the absence of knowledge about the environment the simplest solution is often the most likely: Wagemans et al. (2012a) suggest that evolution may well have built a surrogate for *likelihood* into the visual system via *simplicity*.

A commitment to *Prägnanz* (or *simplicity*) is perhaps the closest to the classical view of Gestalt as employing innate laws of perceptual organisation.

But even here, theoretical abstraction has to give way to empirical reality. As Wagemans et al. (2012a) observe, Gestalt principles are no longer thought of as simply pre-attentive grouping principles, but operate instead at multiple levels and can be heavily influenced by past experience.

By contrast, a third interpretation of Gestalt brings its grouping principles closer to Intentionality. Koenderink (2010) suggests that 'perceptual organisation' is a process of attributing *subjective meaning* to a scene; so rather than asking which is the *statistically most likely* interpretation, or even the *simplest* one, we ask which is the *most rational*: 'There is simply no way to "transform" mere structure into meaning, you—as *perceiver*—have to *supply* it.'

4. Intentionality: Indeed Albertazzi et al. (2010) argue that the insights of early-twentieth century Gestalt Psychology derive from a deeper truth articulated in the late-nineteenth century by Brentano (1874), namely the *act of intentional reference*, according to which:

> ...the structure of a process of seeing, thinking, judging, and so on is that of a *dynamic whole endowed with parts* in which the parts are nonindependent items, and that this act can give rise to relatively different outputs based on *subjective completions*...

But Albertazzi et al.'s aspirations for Intentionality go further still:

> The linking theme is the foundational role of perception as the origin of every potential level of signification, from the most concrete to the most abstract (Arnheim 1969), and a particularly strong interest in the *qualitative* aspects of experience, for within these lie the clues to a richer semantic theory of information.

Albertazzi et al. illustrate their point that vision ought to be understood as much in terms of *qualities* as the *quantities* of geometry and scale, with the example of aesthetic properties: they argue that we *see* aesthetic properties, and yet there is no place for aesthetic properties amongst the traditional primary (geometry and scale) and secondary (colour) qualities of vision. Albertazzi et al. argue that what is required to accommodate such properties is, instead, 'a theory of perception that sees qualitative phenomena and the subjective operations of the observer as foundational.'

But why is this of any interest to us? After all, isn't stereopsis simply concerned with the *quantities* of scale and geometry that are already admitted? Not so, according to Vishwanath (2010), who argues that a 2D image can effectively convey the three-dimensional properties of the scene *without* stereopsis. We will explore in Chap. 3 whether this position is sustainable, but Vishwanath takes this as evidence that stereopsis must reflect *something more* than the three-dimensional properties of the scene, specifically a *quality* of vision that reflects a more *subjectively meaningful* layer of depth, namely 'the depth used to guide motor function' (Volcic et al. 2014). But what does this mean?

Well, to understand Vishwanath's account of *stereopsis* we first have to understand his account of the *surfaces* of objects. Vishwanath (2010) argues that the surfaces of objects ought to be understood as invitations to interact with the world; specifically, they are *anticipatory* structures: the presentation of complex motor *plans*. But how are we to test the validity of such plans? Will we successfully interact with the world if we follow them? Or will we fail? The obvious answer is to simply to try them and see: some motor plans will result in success, others in failure. But from an evolutionary perspective this has huge costs, with every failure being potentially fatal. And this is where a role for stereopsis as a subjective quality of visual experience begins to emerge for Vishwanath:

> Conveniently, my perceptual system has given me a way of being implicitly weary of putting all faith in the 3D presentation before me: by modulating the perceived plastic quality of that 3D presentation.

And so stereopsis becomes the means by which the visual system conveys the *reliability* of the complex motor plans that surfaces represent. Specifically, whilst Vishwanath suggests that our impression of the geometry of an object is largely accurate (even from a 2D image, where stereopsis is absent), what is required to successfully interact with an object is that we not only have (a) its geometry, but also (b) access to reliable egocentric distance information by which to scale its geometry: Is it a small object up close or a large object far away? Consequently, for Vishwanath, stereopsis is the visual system's way of communicating to the observer the precision with which it is able to scale the geometry of the scene or object. Whether this is a sustainable position is explored in Chap. 3.

REFERENCES

Albertazzi, L., van Tonder, G. J., & Vishwanath, D. (2010). *Perception beyond inference: The informational content of visual processes.* Cambridge, MA: MIT Press.

al-Haytham. (c.1028–1038). Book of Optics. In A. M. Smith (trans. & ed.) (2001). *Alhacen's theory of visual perception, volume two: English translation, transactions of the american philosophical society* (Vol. 91, Part 5). Philadelphia: American Philosophical Society.

Ames, A., Jr. (1925a). The illusion of depth from single pictures. *Journal of the Optical Society of America, 10*(2), 137–148.

Ames, A., Jr. (1925b). Depth in pictorial art. *The Art Bulletin, 8*(1), 4–24.

Ames, A., Jr. (1951). Visual perception and the rotating trapezoidal window. *Psychological Monographs, 65*(7), 324.

Ames, A., Jr. (1955). *An interpretative manual: The nature of our perceptions, prehensions, and behavior.* For the Demonstrations in the Psychology Research Center, Princeton University. Princeton, NJ: Princeton University Press.

Arnheim, R. (1969). *Visual Thinking.* Berkeley and Los Angeles, CA: University of California Press.

Aschenbrenner, C. (1954). Problems in getting information into and out of air photographs. *Photogrammetric Engineering, 20*(3), 398–401.

Banks, E. C. (2001). Ernst Mach and the episode of the monocular depth sensations. *Journal of the History of the Behavioural Sciences, 37*(4), 327–348.

Barlow, H. B., Blakemore, C., & Pettigrew, J. D. (1967). The neural mechanism of binocular depth perception. *Journal of Physiology, 193,* 327–342.

Barry, S. (2009). *Fixing my gaze: A scientist's journey into seeing in three dimensions.* New York: Basic Books.

Berkeley, G. (1709). *An essay towards a new theory of vision.* Dublin: Printed by Aaron Rhames, at the back of Dick's Coffee-House, for Jeremy Pepyat, bookseller in Skinner-Row.

Bishop, P. O., & Pettigrew, J. D. (1986). Neural mechanisms of binocular vision. *Vision Research, 26*(9), 1587–1600.

Brentano, F. (1874). *Psychology from an empirical standpoint* (A.C. Rancurello, D.B. Terrell, & L. McAlister Trans.) (1973). London: Routledge.

Bruner, J. S., & Goodman, C. C. (1947). Value and needs as organizing factors in perception. *Journal of Abnormal and Social Psychology, 42,* 33–44.

Burge, J., Fowlkes, C. C., & Banks, M. S. (2010). Natural-scene statistics predict how the figure-ground cue of convexity affects human depth perception. *Journal of Neuroscience, 30,* 7269–7280.

Cajal, S. R. (1904). Textura del Sistema Nervioso del Hombre y los Vertebrados (N. Swanson & L. W. Swanson Trans.) (1995). In *Histology of the nervous system of man and vertebrates.* Oxford: Oxford University Press.

Carr, H. A. (1935). *An introduction to space perception.* New York: Longmans, Green, & Co.

Cavanagh, P. (2011). Visual cognition. *Vision Research, 51*(13), 1538–1551.

Chen, C. C., & Tyler, C. W. (2015). Shading beats binocular disparity in depth from luminance gradients: Evidence against a maximum likelihood principle for Cue combination. *PLoS ONE, 10*(8), e0132658.

Claparède, E. (1904). Stéréoscopie monoculaire paradoxale. *Annales d'Oculistique, 132,* 465–466.

Cumming, B. G., & Parker, A. J. (1999). Binocular neurons in V1 of awake monkeys are selective for absolute, not relative, disparity. *Journal of Neuroscience, 19,* 5602–5618.

Domini, F., & Caudek, C. (2011). Combining Image Signals before Three-Dimensional Reconstruction: The Intrinsic Constraint Model of Cue Integration. In J. Trommershäuser, K. Körding, & M. Landy (Eds.), *Sensory cue integration.* Oxford: Oxford University Press.

Eby, D. W., & Braunstein, M. L. (1995). The perceptual flattening of three-dimensional scenes enclosed by a frame. *Perception, 24*(9), 981–993.

Engel, A. K., Friston, K. J., & Kragic, D. (2016). *The pragmatic turn: Toward action-oriented views in cognitive science.* Cambridge, MA: MIT Press.

Engel, A. K., Maye, A., Kurthen, M., & König, P. (2013). Where's the action? The pragmatic turn in cognitive science. *Trends in Cognitive Science, 17*(5), 202–209.

Enright, J. T. (1989). *Paradoxical monocular stereopsis and perspective vergence.* NASA, Ames Research Center, Spatial Displays and Spatial Instruments, N90-22922.

Fine, I., Wade, A. R., Brewer, A. A., May, M. G., Goodman, D. F., Boynton, G. M., et al. (2003). Long-term deprivation affects visual perception and cortex. *Nature Neuroscience, 6*(9), 915–916.

Firestone, C., & Scholl, B. J. (2016a). Cognition does not affect perception: Evaluating the evidence for 'top-down' effects. *Behavioral and Brain Sciences, 39,* 1–19.

Firestone, C., & Scholl, B. J. (2016b). Seeing and thinking: Foundational issues and empirical horizons. *Behavioral and Brain Sciences, 39,* 53–67.

Gabor, D. (1960). Three-dimensional cinema. *New Scientist,* 14th July 1960, 141.

Gibson, J. J. (1947). *Motion picture testing and research.* Research Reports, Report No. 7, Army Air Forces Aviation Psychology Program.

Gibson, J. J. (1950). *The perception of the visual world.* Boston: Houghton Mifflin.

Gibson, J. J. (1954). A Theory of pictorial perception. *Audio-Visual Communication Review, 1,* 3.

Gibson, J. J. (1960). Pictures, Perspective, and Perception. *Daedalus, 89,* 216.

Gibson, J. J. (1966). *The senses considered as perceptual systems*. Boston: Houghton Mifflin.

Gibson, J. J. (1971). The information available in pictures. *Leonardo, 4,* 27–35.

Gibson, J. J. (1979). *The ecological approach to visual perception*. Boston: Houghton Mifflin.

Gilchrist, A. (2006). *Seeing black and white*. Oxford: Oxford University Press.

Gombrich, E. H. (1960). *Art and illusion: A study in the psychology of pictorial representation*. London: Phaidon.

Gregory, R. L. (1966). *Eye and brain: The psychology of seeing*. London: Weidenfeld & Nicolson.

Gregory, R. L. (1970). *The intelligent eye*. London: Weidenfeld & Nicolson.

Gregory, R. L. (1980). Perception as hypothesis. *Philosophical Transactions of the Royal Society B, 290*(1038), 181–197.

Gregory, R. L. (2004). The blind leading the sighted. *Nature, 430,* 1.

Gregory, R. L., & Wallace, J. G. (1963). *Recovery from early blindness: A case study*. Experimental Psychology Society Monograph No. 2. Cambridge: Heffer.

Helmholtz, H. von. (1866). Physiological Optics, Vol. 3. In J. P. C. Southall (Trans. & ed.) (1925). *Treatise on Physiological Optics*. New York: Dover.

Hering, E. (1865). Ueber stereoskopisches Sehen. *Verhandlungen des naturhistorisch-medizinischen Vereins zu Heidelberg, 3,* 8–11.

Hibbard, P. (2008). Can appearance be so deceptive? Representationalism and binocular vision. *Spatial Vision, 21*(6), 549–559.

Holt, E. (1904). Die von M. von Rohr gegebene Theorie des Veranten, eines Apparats zur Richtigen Betrachtung von Photographien by E. Wandersleb; The Verant, a New Instrument for Viewing Photographs from the Correct Standpoint by M. von Rohr; Der Verant, ein Apparat zum Betrachten von Photogrammen in Richtigen Abstande by A. Köhler. *The Journal of Philosophy, Psychology and Scientific Methods, 1*(20), 552–553.

Julesz, B. (1960). Binocular depth perception of computer-generated patterns. *Bell Labs Technical Journal, 39,* 1125–1162.

Kanizsa, G. (1955). Margini quasi-percettivi in campi con stimolazione omogenea. *Rivista di Psicologia, 49*(1), 7–30.

Knill, D. C. (2007). Learning Bayesian priors for depth perception. *Journal of Vision, 7*(8), 13.

Knill, D. C., & Richards, W. (1996). *Perception as Bayesian inference*. Cambridge: Cambridge University Press.

Koenderink, J. J. (2010). Vision and information. In L. Albertazzi, J. van Tonder, & D. Vishwanath (Eds.), *Perception beyond inference: The informational content of visual processes*. Cambridge, MA: MIT Press.

Koenderink, J. J., van Doorn, A. J., & Kappers, A. M. L. (1994). On so-called paradoxical monocular stereoscopy. *Perception, 23,* 583–594.

Koenderink, J. J., van Doorn, A., Albertazzi, L., & Wagemans, J. (2015a). Relief articulation techniques. *Art & Perception, 3*(2), 151–171.

Koenderink, J. J., van Doorn, A., & Wagemans, J. (2015b). Part and whole in pictorial relief. *i-Perception, 6*(6), 1–21.

Koenderink, J. J., Wijntjes, M. W. A., & van Doorn, A. J. (2013). Zograscopic viewing. *i-Perception, 4*(3), 192–206.

Kosslyn, S. M. (2006). You can play 20 questions with nature and win: Categorical versus coordinate spatial relations as a case study. *Neuropsychologia, 44*(9), 1519–1523.

Landy, M., Banks, M., & Knill, D. (2011). Ideal-observer models of cue integration. In J. Trommershäuser, K. Körding, & M. Landy (Eds.), *Sensory cue integration*. Oxford: Oxford University Press.

Landy, M. S., Maloney, L. T., Johnston, E. B., & Young, M. (1995). Measurement and modeling of depth cue combination: In defense of weak fusion. *Vision Research, 35,* 389–412.

Langer, M. S., & Bülthoff, H. H. (2000). Depth discrimination from shading under diffuse lighting. *Perception, 29*(6), 649–660.

Livingstone, M. (2002). *Vision and art: The biology of seeing.* New York: Abrams.

Mach, E. (1868). *Beobachtungen über monoculare Stereoscopie.* Sitzungsberichte der kaiserlichen Akademie, mathematische-naturwissenschaftliche Klasse, Wien, 58, 731–736.

Mach, E. (1886). *The analysis of sensations and the relation of the physical to the psychical* (C. M. Williams Trans.) (1959). New York: Dover.

Mather, G. (1989). The role of subjective contours in capture of stereopsis. *Vision Research, 29,* 143–146.

Miller, G. A. (2003). The cognitive revolution: A historical perspective. *Trends in Cognitive Science, 7*(3), 141–144.

Morgan, M. J., & Watt, R. J. (1982). Mechanisms of interpolation in human spatial vision. *Nature, 299,* 553–555.

Münsterberg, H. (1904). Perception of distance. *Journal of Philosophy, Psychology and Scientific Methods, 1*(23), 617–623.

Nakayama, K. (1999). Mid-level vision. In R. A. Wilson & F. C. Keil (Eds.), *The MIT encyclopaedia of the cognitive sciences.* Cambridge, MA: MIT Press.

Nakayama, K. (2005). Resolving border disputes in midlevel vision. *Neuron, 47,* 5–8.

Nakayama, K., Shimojo, S., & Silverman, G. H. (1989). Stereoscopic depth: Its relation to image segmentation, grouping, and the recognition of occluded objects. *Perception, 18*(1), 55–68.

Neisser, U. (1967). *Cognitive Psychology.* New York: Appleton-Century-Crofts.

Nguyen, J., Majmudar, U. V., Ravaliya, J. H., Papathomas, T. V., & Torres, E. B. (2016). Automatically characterizing sensory-motor patterns underlying reach-to-grasp movements on a physical depth inversion illusion. *Frontiers in Neuroscience, 9,* 694.

Nikara, T., Bishop, P. O., & Pettigrew, J. D. (1968). Analysis of retinal correspondence by studying receptive fields of binocular single units in cat striate cortex. *Experimental Brain Research, 6*, 353–372.

Ogle, K. (1950). *Researches in Binocular Vision.* Philadelphia, PA: Saunders.

Ogle, K. (1954). On stereoscopic depth perception. *Journal of Experimental Psychology, 48*(4), 225–233.

Ogle, K. (1959). The theory of stereoscopic vision. In S. Koch (ed.). *Psychology: A study of a science, vol. I, sensory, perceptual and physiological formulations* (362–394). New York: McGraw Hill.

Parker, A. J. (2007). Binocular depth perception and the cerebral cortex. *Nature Reviews Neuroscience, 8*(5), 379–391.

Parker, A. J. (2016). Vision in our three-dimensional world. *Philosophical Transactions of the Royal Society B, 371*(1697), 20150251.

Peacocke, C. (1983). *Sense and content: Experience, thought, and their relations.* Oxford: Oxford University Press.

Pentland, A. (1982). The perception of shape from shading. Proceedings of the OSA Annual Meeting, Oct. 18–22, Tuscon, AZ.

Pettigrew, J. D. (1965). *Binocular interaction on single units of the striate cortex of the cat.* Thesis, University of Sydney.

Poston, T., & Stewart, I. (1978). Nonlinear modeling of multistable perception. *Systems Research and Behavioural Science, 23*(4), 318–334.

Pylyshyn, Z. (1999). Is vision continuous with cognition? The case for cognitive impenetrability of visual perception. *Behavioral and Brain Sciences, 22*(3), 341–365.

Qiu, F. T., & von der Heydt, R. (2005). Figure and ground in the visual cortex: V2 combines stereoscopic cues with gestalt rules. *Neuron, 47*(1), 155–166.

Ramachandran, V. S. (1986). Capture of stereopsis and apparent motion by illusory contours. *Perception & Psychophysics, 39*(5), 361–373.

Ramachandran, V. S. (1988). Perception of shape from shading. *Nature, 331,* 163–166.

Ramachandran, V. S. (2006). Foreword. In A. Gilchrist (Ed.), *Seeing black and white.* Oxford: Oxford University Press.

Ramachandran, V. S., & Cavanagh, P. (1985). Subjective contours capture stereopsis. *Nature, 317,* 527–530.

Ramachandran, V. S., & Rogers-Ramachandran, D. (2009). Two eyes, two views: Your brain and depth perception. *Scientific American,* 1st Sept. 2009.

Rock, I., & Palmer, S. (1990). The legacy of Gestalt psychology. *Scientific American, 263*(6), 84–90.

Rogers, B., & Gyani, A. (2010). Binocular disparities, motion parallax, and geometric perspective in Patrick Hughes's 'reverspectives': Theoretical analysis and empirical findings. *Perception, 39*(3), 330–348.

Rosas, P., & Wichmann, F. A. (2011). Cue combination: Beyond optimality. In J. Trommershäuser, K. Körding, & M. Landy (Eds.), *Sensory cue integration*. Oxford: Oxford University Press.

Sacks, O. (2006). Stereo Sue. In O. Sacks (Ed.), *The minds eye (2010)*. New York: Picador.

Sacks, O. (2010). Persistence of vision: A journal. In O. Sacks (Ed.), *The minds eye*. New York: Picador.

Scarfe, P., & Hibbard, P. B. (2011). Statistically optimal integration of biased sensory estimates. *Journal of Vision, 11*(7), 1–17.

Schlosberg, H. (1941). Stereoscopic depth from single pictures. *The American Journal of Psychology, 54*(4), 601–605.

Seydell, A., Knill, D. C., & Trommershäuser, J. (2011). Priors and learning in cue integration. In J. Trommershäuser, K. Körding, & M. Landy (Eds.), *Sensory cue integration*. Oxford: Oxford University Press.

Sprague, W. W., Cooper, E. A., Reissier, S., Yellapragada, B., & Banks, M. S. (2016). The natural statistics of blur. *Journal of Vision, 16*(10), 23, 1–27.

Strawson, P. F. (1979). Perception and its objects. In G. F. Macdonald (ed.), *Perception and identity: Essays presented to A. J. Ayer, with his replies* (41–60). London: Macmillan.

Tausch, R. (1953). Die beidäugige Raumwahrnehmung. *Zeitschrift fur Experimentelle und Angewandte Psychologie, 3*, 394–421.

Todd, J. T., Egan, E. J., & Kallie, C. S. (2015). The darker-is-deeper heuristic for the perception of 3D shape from shading: Is it perceptually or ecologically valid? *Journal of Vision, 15*(5), 2.

Trommershäuser, J., Körding, K., & Landy, M. S. (2011). *Sensory cue integration*. Oxford: Oxford University Press.

Turner, R. S. (1994). *In the eye's mind: Vision and the Helmholtz-Hering controversy*. Princeton, NJ: Princeton University Press.

Tye, M. (1993). Blindsight, the absent qualia hypothesis, and the mystery of consciousness. *Royal Institute of Philosophy Supplement, 34*, 19–40.

Tyler, C. W. (1998). Diffuse illumination as a default assumption for shape-from-shading in the absence of shadows. *Journal of Imaging Science and Technology, 42*(4), 319–325.

Tyler, C. W. (2004). Theory of texture discrimination of based on higher-order perturbations in individual texture samples. *Vision Research, 44*(18), 2179–2186.

Tyler, C. W. (2006). Spatial form as inherently three dimensional. In M. R. M. Jenkin & L. R. Harris (Eds.), *Seeing spatial form*. Oxford: Oxford University Press.

Vishwanath, D. (2005). The epistemological status of vision science and its implications for design. *Axiomathes, 15*(3), 399–486.

Vishwanath, D. (2010). Visual information in surface and depth perception: Reconciling pictures and reality. In L. Albertazzi, G. van Tonder, & D. Vishwanath (Eds.), *Perception beyond inference: The informational content of visual processes*. Cambridge, MA: MIT Press.

Vishwanath, D. (2016). Induction of monocular stereopsis by altering focus distance: A test of Ames's hypothesis. *i-Perception, 7*(2), 1–5.

Vishwanath, D., & Hibbard, P. B. (2013). Seeing in 3D with just one eye: Stereopsis without binocular vision. *Psychological Science, 24*(9), 1673–1685.

Volcic, R., Vishwanath, D., & Domini, F. (2014). Reaching into pictorial spaces. *Proceedings of SPIE, 9014: Human Vision and Electronic Imaging XIX.*

von der Heydt, R. (2015). Figure-ground organization and the emergence of proto-objects in the visual cortex. *Frontiers in Psychology., 6,* 1695.

von Rohr, M. (1903). The verant, a new instrument for viewing photographs from the correct standpoint. *The Photographic Journal, 43,* 279–290.

Wagemans, J., Elder, J. H., Kubovy, M., Palmer, S. E., Peterson, M. A., Singh, M., et al. (2012a). A century of Gestalt psychology in visual perception: I. Perceptual grouping and figure-ground organization. *Psychological Bulletin, 138*(6), 1172–1217.

Wagemans, J., Feldman, J., Gepshtein, S., Kimchi, R., Pomerantz, J. R., van der Helm, P., et al. (2012b). A century of Gestalt psychology in visual perception: II. Conceptual and theoretical foundations. *Psychological Bulletin, 138*(6), 1218–1252.

Wagemans, J., van Doorn, A. J., & Koenderink, J. J. (2010). The shading cue in context. *i-Perception, 1,* 159–177.

Wallis, G., & Ringelhan, S. (2013). The dynamics of perceptual rivalry in bistable and tristable perception. *Journal of Vision, 13*(2): 24, 1–21.

Wertheimer, M. (1924). Ueber Gestalttheorie. Lecture before the Kant Gesellschaft.

Westheimer, G. (1994). The Ferrier Lecture, 1992. Seeing depth with two eyes: stereopsis. *Proceedings of the Royal Society B, 257*(1349), 205–214.

Westheimer, G. (2013). Clinical evaluation of stereopsis. *Vision Research, 20* Sept. 2013, 38–42.

Wheatstone, C. (1838). Contributions to the theory of vision—Part the first, on some remarkable, and hitherto unobserved, phenomena of binocular vision. *Philosophical Transactions of the Royal Society, 128,* 371–394.

Wijntjes, M. W. A., Füzy, A., Verheij, M. E. S., Deetman, T., & Pont, S. C. (2016). The synoptic art experience. *Art & Perception, 4*(1–2), 73–105.

Wilde, K. (1950). Der Punktreiheneffekt und die Rolle der binokularen Querdisparation beim Tiefensehen. *Psychologische Forschung, 23,* 223–262.

Zhou, H., Friedman, H. S., & von der Heydt, R. (2000). Coding of border ownership in Monkey visual cortex. *Journal of Neuroscience, 20*(17), 6594–6611.

CHAPTER 2

Stereopsis in the Presence of Binocular Disparity

Abstract The strongest evidence that pictorial cues contribute to stereopsis is the fact that the visual system appears to integrate both pictorial cues (such as perspective and shading) and optical cues (such as binocular disparity) into a single coherent percept. Indeed, when these sources of information are slightly in conflict the visual system appears to construct an entirely new object that is not specified by any of the individual sources of information. But in this chapter I question whether what we experience in this context is really an *integrated percept*, as opposed to an *integrated judgment*, and I suggest experimental strategies that might enable us to distinguish between these two interpretations.

Keywords Cue conflict · Cue combination · Cue Integration Mandatory fusion · Hollow-Face illusion · Reverspective

In Chap. 1, we saw how one of the strongest arguments against a purely optical account of stereopsis was the fact that pictorial cues appear to be able to modify the depth specified by binocular disparity. This effect was well documented in the mid-twentieth century by Ames (1951, 1955)

The original version of this chapter was revised: Post-publication corrections have been incorporated. The erratum to this chapter is available at https://doi.org/10.1007/978-3-319-66293-0_5

© The Author(s) 2017
P. Linton, *The Perception and Cognition of Visual Space*,
DOI 10.1007/978-3-319-66293-0_2

35

and Ogle (1959), and in the 1970s–1980s by Gregory's (1970) hollow-face illusion and Hughes' Reverspectives. But in the mid-1990s the ability of the visual system to reconcile conflicting sources of information into a single coherent percept became the *organising principle* of perception (see Landy et al. 1995; Knill and Richards 1996). You might reasonably wonder why? After all, the human visual system evolved in response to a cue-consistent real world rather than the artificially induced cue conflicts of the laboratory. But as Hillis et al. (2002) explain, contemporary articulations of Cue Integration start from the premise that every depth cue is subject to two sources of potential error, namely *bias* (inaccuracy) and *random noise* (imprecision), and this explains why estimates of the same property from different cues are liable to differ.

Furthermore, whilst some authors have explored potential *bias* (see Domini and Caudek 2011; Scarfe and Hibbard 2011), the majority of the Cue Integration literature proceeds on the basis that *random noise* is the most egregious concern. Hillis et al. (2002) are therefore typical when they assume that the visual system is well calibrated, so that signals will *on average* agree with one another. Instead, they assume that the source of any discrepancy is the random error that all measurements are subject to and which can be modelled by a Gaussian distribution. And it is upon this basis that Cue Integration has generally embraced a Bayesian weighted average as the most appropriate model of cue combination.

This Bayesian model of Cue Integration is typically evaluated by introducing conflicts between various depth cues and seeing if the visual system responds as expected. But one might reasonably question whether the cue conflicts employed in these experimental studies actually reflect a concern for random noise? For instance, it is hard to maintain that Ernst et al. (2000) (where 0° texture was pitted against 30° binocular disparity, and vice versa) or Hillis et al. (2002) (where +20° texture was pitted against −20° binocular disparity, and vice versa) merely modelled random noise in the visual system. Indeed, as Hillis et al. (2002) readily admit 'such combinations rarely occur in the natural environment'; a point that Landy et al. (2011) reiterate: 'one might argue that the artificial stimuli create cue conflicts that exceed those experienced under natural conditions…'.

Nonetheless, even in the context of these artificially accentuated cue conflicts, the visual system often appears to integrate depth cues in a linear fashion. So even if the Bayesian justification for these studies begins to look questionable, their empirical findings, and especially the method used to procure them (the 'perturbation analysis' of inducing small cue conflicts: see Maloney and Landy 1989; Landy et al. 1991; Young et al. 1993), have become standard in the literature. Indeed criticism of Bayesian Cue Integration typically comes from those whose empirical findings in cue-conflict experiments are not consistent with a weighted average: see Domini and Caudek (2011) for an overview; and in particular Todd and Norman (2003), Likova and Tyler (2003), Vishwanath and Domini (2013), Vishwanath and Hibbard (2013), and Chen and Tyler (2015) for our present purposes.

But underlying this debate is the common assumption that the cue-conflict stimuli in these experiments really are integrated into a *single coherent percept*. This is true for those who advance a Bayesian account of Cue Integration, those who embrace an alternative conception of Cue Integration (see Domini and Caudek 2011; Tyler 2004), and even those who reject Cue Integration altogether (see Vishwanath 2005; Albertazzi et al. 2010; Koenderink 2010). For instance, although Vishwanath (2005) rejects Cue Integration, he nonetheless maintains that 'cue-conflict stimuli are ideal for studying how co-calibration across sensory measurements is maintained: a calibration process that is designed to remove detected conflicts when possible.' By contrast, the purpose of this chapter is to challenge the assumption that cue-conflicts are really eradicated at the level of *perception*. So whilst critics of Bayesian Cue Integration may challenge *how* these sources of information are perceptually integrated, I am asking the logically prior question of *if* they are perceptually integrated in the first place? But if they are not *perceptually* integrated, then what is the alternative? Well, the literature appears to draw a false dichotomy between (a) a *single integrated percept* and (b) *strategic decision-making*. For instance, in the context of vision and touch, Gepshtein et al. (2005) find evidence of subjects relying upon both sources of information, and ask: 'Do the results manifest a unified multi-modal percept?' And they admit that their results are silent between two competing interpretations:

> The improvement in precision observed in the inter-modality experiment could in principle result from a perceptual process or a decision strategy.

And Gepshtein et al. clarify what each interpretation would entail

> By the former, we mean that the observer's judgements are based on a unified multi-modal estimate resulting from the weighted combination of visual and haptic signals (Hillis, Ernst, Banks, & Landy, 2002).

> By the latter, we mean that the observer's decision is based solely on comparing (and weighting appropriately) the two unimodal signals without actually combining them into a unified percept. That is, the information could still be used optimally, but without the percept of a single object.

Ultimately Gepshtein et al. conclude

> Our study cannot distinguish between these two possibilities…

Similarly, when Todd and Norman (2003) observed that their subjects gave inconsistent evaluations of the stimuli depending upon how the stimuli were presented, they concluded that a strategic element must be at play:

> The incompatibility of the objective data with the observers' phenomenal impressions provide strong evidence that there was a strategic component of their responses that was not based entirely on their conscious perceptions.

But I would argue that there is a third possibility, namely (as I outlined in Chap. 1) that rather than engaging in *conscious deliberation*, the subjects in Todd and Norman (2003) might simply be influenced by an *automatic* and *involuntary* cognitive process that operates *after* perception but *prior* to conscious deliberation. For instance, the attribution of meaning to words is not a *perceptual* process (it does not affect the visual appearance of the words themselves) and yet clearly operates *preconsciously* (we do not have to consciously attribute meaning to the words). But if a preconscious cognitive process can account for the attribution of meaning to words, why not the attribution of meaning to depth cues? Under this account the subjects' evaluations in Todd & Norman (2003) are not strategic but based upon an *integrated evaluative judgement* that had already occurred earlier in the cognitive chain. So the question we need to ask is whether cue-conflict stimuli really provide evidence for an *integrated percept* or merely an *unconscious post-perceptual integrated judgement*?

1   DOES CUE INTEGRATION CLAIM PERCEPTUAL FUSION?

One commentator has suggested that I have misinterpreted Cue Integration theory: they argue that Cue Integration is solely concerned with *performance*, rather than the basis of that performance, and so remains agnostic between these three different interpretations (perception, unconscious judgement, conscious decision strategy). I disagree for the following four reasons:

1. First, Cue Integration theorists are clearly cognisant of this argument. For instance, Held et al. (2012b) admit that in depth perception studies a conscious decision strategy based on a 2D interpretation of the cues is essentially always available, but generally unacknowledged. But what is interesting is that in their study of depth from defocus blur, Held et al. (2012b) reject this 2D interpretation by appealing to their subjects' visual experience: 'An important clue is subjects' phenomenology'. They asked their subjects whether they were relying on perceived depth or merely (as might be possible for defocus blur) a 2D inference and found that only one out of their four subjects relied on a 2D strategy, and even then only rarely. Consequently, they concluded that their findings must reflect *perception*.

Indeed, appeals to the subjects' own visual experience are made not only to confirm data that appear to be consistent with Cue Integration (such as Held et al. 2012b) but also, more controversially, to discount data that appear to contradict Cue Integration. For instance, when the subjects in Hillis et al. (2002) were able to discriminate stimuli that ought to be indiscriminable so far as Cue Integration is concerned: 'The participants' phenomenology was informative.' The subjects reported that the stimulus introduced 2D texture distortions that enabled them to discriminate between the stimuli, enabling Hillis et al. to maintain that the 3D cues to slant were truly fused, and that subjects only had access to a single depth percept, in spite of their contradictory performance.

2. Second, the very rationale of Cue Integration suggests that integration must *perceptual*: if the purpose of Cue Integration is to reduce the impact of random system noise by averaging across various noisy cues, why would the visual system give us direct access (via perception) to one of these noisy cues? The implication, as Hillis et al. (2002) explain, is that Cue Integration must not only have a *positive* dimension (improved performance when reliance on two or more cues would be beneficial),

but also a *negative* dimension (reduced performance when reliance on one cue alone would be beneficial). Hillis et al. (2002) term this negative dimension of Cue Integration *mandatory fusion*; specifically, a *loss of access* to individual depth cues: the visual system specifies a *single* depth estimate, which is beneficial from an evolutionary perspective (in a cue-consistent world, discrepancies are more likely to come from the visual system's inconsistent measurements), but which leads to detrimental performance in response to artificially contrived cue conflicts in the laboratory.

3. Third, *mandatory fusion* is an inevitable consequence of Cue Integration for another, more immediate, reason; namely, how Cue Integration conceives of the depth estimates from individual cues: It doesn't treat the sensory data as a *specific* estimate of depth, but rather as the basis for a probability distribution (a 'likelihood function') which plots the probability of receiving *this* sensory data from a variety of different potential depth values whose signal has been corrupted by noise. Consequently, there is no possibility that perception reflects *the* estimate from one specific cue, since there is no one specific estimate, only a set of probabilities.

4. Fourth, the final reason that we know Cue Integration is committed to *perceptual* integration is that many of its most startling claims are articulated as claims about *visual experience* rather than *performance*. Consider, for instance, Ernst et al.'s (2000) paper: 'Touch Can Change Visual Slant Perception'. Ernst et al. went beyond merely an observation about *performance*, namely that touch feedback can affect the weight given to various sources of visual information, to an observation about *visual experience*, namely that touch can change the slant that is *seen*. Indeed, as the title of their paper illustrates, it was this claim about *visual experience*, rather than the improved cross-modal *performance*, that proved to be the central message of their paper.

The claim that touch can influence the slant that is *seen* has been thrown into doubt by the subsequent literature. For instance, as the title of their paper suggests, Hillis et al. (2002) found 'mandatory fusion within, but not between, senses'. Nonetheless, even if we stick to Cue Integration within vision itself, mandatory fusion has quite profound implications for our visual experience: as Hillis et al. explain, an appropriately calibrated high-texture-low-disparity stimulus and low-texture-high-disparity stimulus should be *perceptually indistinguishable.*

Indeed, Hillis et al. develop this point with an analogy from the colour literature, namely *metamers*: 'composite stimuli that cannot be discriminated even though their constituents can be'. So just as red and green light added together is subjectively indistinguishable from yellow light, cue-conflict stimuli can be subjectively indistinguishable even though had their disparity or texture been presented in isolation you would be able to differentiate them. Indeed, Hillis et al.'s *metamer* analysis had such a profound effect on the literature that within two years, it was legitimate for Ernst and Bülthoff (2004) to simply assume mandatory fusion as an initial premise rather than a conclusion that had to be argued for.

2   DOES CUE INTEGRATION DEMONSTRATE PERCEPTUAL FUSION?

But how do Hillis et al. (2002) justify their claim that Cue Integration produces *metamers*? I.e. that an appropriately calibrated high-texture-low-disparity stimulus and a low-texture-high-disparity stimulus are *perceptually* indistinguishable?

First, Hillis et al. take a cue-consistent stimulus (with the same slant specified by texture and disparity) and introduce a cue conflict by varying the stimulus along one of two dimensions until the subject is able to identify the altered stimulus (by picking the odd-one-out when the altered stimulus and two unaltered stimuli are shown in succession). The results of this preliminary study were then used to mark out each subject's subjective thresholds for changes in texture and disparity (the parallel lines in Fig. 1, left), and the question was whether altering the stimulus along *both* dimensions at the same time could *improve performance* (with subjects noticing two complementary sub-threshold changes in disparity and texture: the *positive* dimension of Cue Integration), or even *worsen performance* (with subjects failing to notice an above-threshold change in the one cue if a change in the opposite direction is made in the other: the *negative* dimension of Cue Integration), as predicted by Hillis et al.'s (2002) model of mandatory fusion?

Figure 2 provides an illustration of the subjects' performance in Hillis et al. (2002). It is unclear how representative these results are, but they suffice for the purposes of our discussion. They do tend to show the

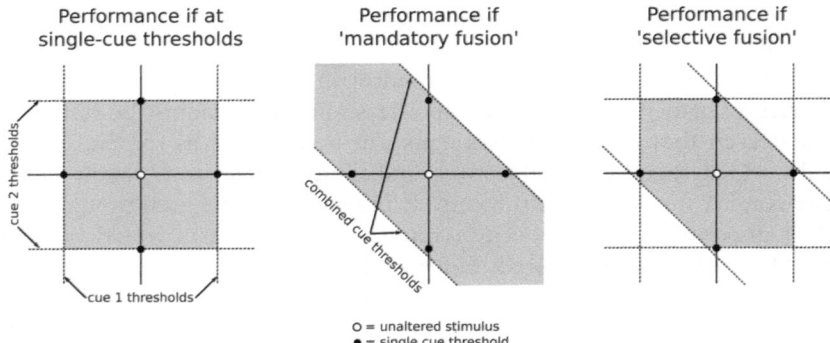

Performance if at single-cue thresholds

Performance if 'mandatory fusion'

Performance if 'selective fusion'

cue 2 thresholds

cue 1 thresholds

combined cue thresholds

O = unaltered stimulus
● = single cue threshold

Fig. 1 Hillis et al.'s (2002) predictions if subjects **a** use single-cue estimates (no Cue Integration), **b** only have access to a combined estimate (mandatory fusion), or **c** experience the benefits of Cue Integration without the costs of mandatory fusion (selective fusion)

predicted *improvement* in performance, but evidence for the predicted *deficit* in performance seems patchy: certainly subject JH seems to be performing close to threshold, as does subject AH on occasion.

Hillis et al. take those instances where there was a performance deficit as evidence of mandatory fusion. But they recognise that mandatory fusion should not be partial, and so try to explain why the predicted deficit wasn't always present. As mentioned above, they suggest that the texture of the stimulus was subject to distortions as disparity was increased, and it was on this basis (rather than 3D slant) that subjects were able to identify the odd-one-out. But there are two concerns with this explanation: The first is that this psychophysical task was chosen as one that would establish mandatory fusion via performance without the need to evaluate experience, so it is concerning to see Hillis et al. using subjective experience to explain away performance that is contrary to their hypothesis. The second is that it is unclear to what extent their explanation maps their results: First, why did it not affect those trials where the predicted performance deficit was found? Second, why does performance from texture distortion (i) coincide almost exactly with the subject's own single-cue thresholds, and (ii) not appear to change significantly as disparity is increased?

Still, we have to explain those instances where the performance deficit predicted by mandatory fusion was present. Is this the clear evidence of mandatory fusion that Hillis et al. suggest?

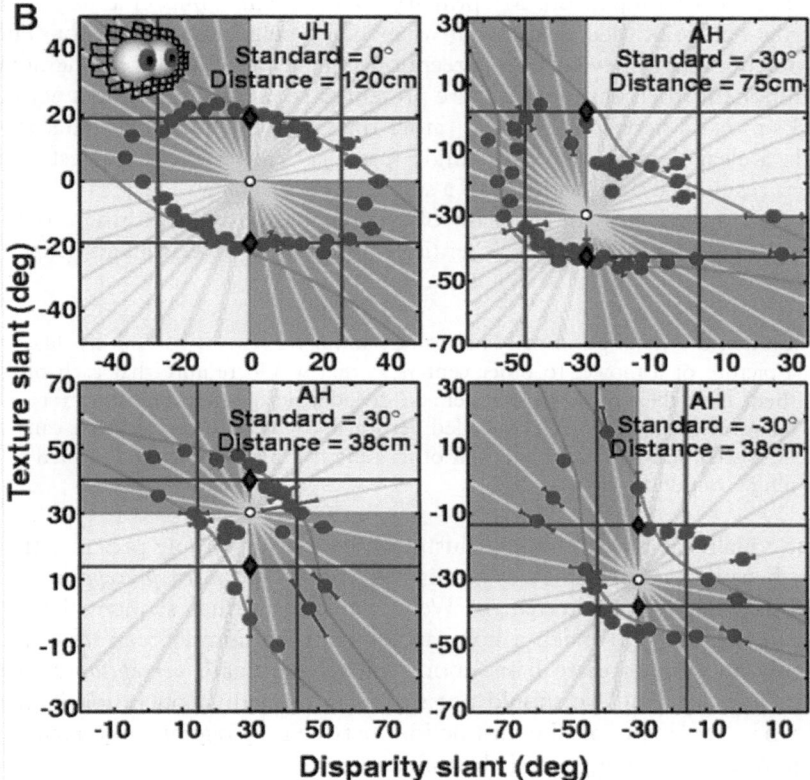

Fig. 2 Illustrative results from two participants in Hillis et al. (2002): The two sets of parallel lines represent the subject's single-cue thresholds. Any point within the rectangle demarcated by the parallel lines represents an improvement over single-cue performance, whilst any point outside the rectangle demarcated by the parallel lines represents a performance deficit relative to single-cue performance. The curves represent Hillis et al.'s model of the optimal combined estimator (Fig. 1, middle) taking into account how the weights assigned vary with texture-specified slant. From Hillis et al. (2002). Combining sensory information: mandatory fusion within, but not between, senses. Science, 298, 1627–1630. © The American Association for the Advancement of Science

1. Direct Comparison: My primary concern with Hillis et al. isn't the partial nature of their results, but what their results are evidence for. The problem is that in evaluating perception Hillis et al. introduce a memory component: their three stimuli are presented sequentially for 1.5 s with a 0.3 s interval between them. So rather than claiming that their data provide a clear demonstration of single fused *percept*, I would suggest that their results are only evidence of a single fused *memory*.

That our *memory* of the slant of a percept is based on a single overall estimate or impression is entirely plausible. As Cavanagh (2011) observes:

> Clearly, the description of visual scene cannot be sent in its entirety, like a picture or a movie, to other centers as that would require that each of them have their own visual system to decode the description. Some very compressed, annotated, or labelled version must be constructed that can be passed on in a format and that other centers – memory, language, planning – can understand.

But equally, as this quotation illustrates, we cannot simply presume that just because *memory* operates in this way, that so too must *perception*.

But what is the alternative? Well, it might be that sequential tasks (asking subjects to make a comparison between stimuli *over time*) are simply an inappropriate basis upon which to evaluate *perception* rather than *memory*. And we should not shy away from this conclusion if that is what the logic of the distinction between *perception* and *memory* requires. That being said, I do think that we can legitimately question why Hillis et al. introduce a 0.3 s interval between their stimuli? If the various stimuli really are *metamers* in the strong sense that Hillis et al. suggest, and is implied by mandatory fusion, then we have to question why this 0.3 s interval is required: for instance, if we wanted to demonstrate that two shades of yellow were subjectively *indistinguishable*, we would simply alternate between them without an interval, so why should 3D depth be any different?

One response, suggested to me by a commentator, is that *motion* might be processed separately from *3D form*: so whilst *3D form* might be indistinguishable, motion detectors may alert the subject that *something* has changed in the stimulus, even though the subject cannot identify what this change was. In one sense, this problem is caused by Hillis et al.'s reliance on *performance* to determine mandatory fusion, leaving subjects free to rely on any means possible to pick the odd-one-out. But do concerns about motion detection from removing the inter-stimulus interval concede too much?

First, the whole point of Cue Integration is that we give *meaning* to noisy cues. If it would be unwise for the visual system to give subjects access to individual noisy cues in the context of *3D form* perception (the mandatory fusion thesis), then why would it make any more sense in the context of *motion*? After all, according to the perturbation analysis, we are meant to be modelling random noise in the visual system. And if motion detectors were triggered every time random noise in the visual system fluctuated, this would be a recipe for evolutionary disaster.

Second, even if subjects could tell the difference between the two stimuli with the interval removed, it could still be a good test so long as we reintroduce an *evaluative* component and asked subjects whether the impression of depth between the two stimuli was *qualitatively* similar? Even if subjects judge the two stimuli to have the same *quantity* of depth, do they really lose nothing (so far as perceptual depth is concerned) as we alternate between them?

In conclusion, we want to avoid reducing our evaluation of visual depth into a change-blindness paradigm, so the choice seems clear: either we test the sequential paradigm *without* an artificially induced interval, or we avoid the sequential paradigm altogether.

2. Subjective Evaluation: At the same time, Hillis et al. (2002) were clearly onto something when they sought to eradicate an *evaluative* element from their task. Asking people whether a stimulus *looks flat, looks slanted,* or *looks bulged,* is as much an evaluative judgement as asking someone if a stimulus *looks square* or *looks symmetrical.* So how do we know that pictorial cues contribute to our perception, rather than merely biasing our evaluation?

Indeed, under my account (where the 3D form of a cue-conflict stimulus is specified solely by its binocular disparity), there are good evolutionary reasons for divorcing our *evaluation* of the scene from our *perception* of it: binocular disparity reduces with distance, but the physical geometry of the scene does not. Consequently, if we wish to use our evaluations as the basis for our interactions with the invariant physical world, we cannot rely too heavily upon our perceptual impression of stereopsis. Indeed, this concern continues to apply (albeit with less force) in the context of linear Cue Integration, where the reduction of binocular disparity with distance still affects the perceived depth of the scene. Nor should we be surprised that our *perception* and *evaluation* of the depth in a scene can come apart: we are quite capable of watching TV at close quarters (e.g. on a laptop screen: 40–50 cm, or even a phone: 30 cm) without the flatness specified by binocular disparity significantly impeding our enjoyment, and this might explain the indifference that the general public has recently shown towards 3D movies.

Like Hillis et al. (2002), the purpose of this chapter is to try and identify 'a true test for the existence of cue fusion.' And like Hillis et al., I am sceptical that relying on subject's evaluative judgements provides that evidence. To see why, consider Ernst et al. (2000). The subjects were shown cue-conflict stimuli with inconsistent slants specified by texture and disparity (Fig. 3). The subjects received touch feedback that was consistent with texture or disparity, and this touch feedback influenced the estimate of slant. But none of the subjects in the experiment noticed that either (a) the slants specified by texture or disparity were different, or (b) that the touch feedback was consistent with one but not the other. So if we were to determine mandatory fusion simply by asking subjects for their subjective impressions we would have at least one *false positive* in this case: as Hillis et al. (2002) have convincingly demonstrated, there is *no* mandatory fusion in the cross-modal context of vision and touch. So the cross-modal integration in Ernst et al. (2000) must simply reflect the subjects' *post-perceptual evaluation* of the stimulus. But in which case, what makes us any more confident that the integration of texture and disparity in the unimodal case of vision is any more *perceptual*? As we learnt from the cross-modal context, the fact that they might *seem* integrated is not enough.

More evidence that we cannot simply delegate this question to subjects' own subjective evaluations comes from Todd and Norman (2003). Todd and Norman asked their subjects to evaluate the depth from (a) a monocular motion display, (b) a static binocular disparity display, and (c) a binocular disparity plus motion display, and found that depth was judged to be highest in the monocular motion display and lowest in the static binocular disparity display, with the binocular disparity plus motion display falling midway between the two. Indeed, all subjects judged the binocular disparity plus motion display to have at least 15% less depth than the monocular motion display. But Todd and Norman asked their subjects to close one eye as they watched the binocular disparity plus motion display, and report whether they saw an increase or a decrease in depth? All the observers reported a significant reduction in the perceived depth, even though closing one eye converts the binocular disparity plus motion display into the monocular motion display that had earlier been evaluated as having 15% more depth.

Todd and Norman correctly conclude that, out of the two results, the direct and immediate comparison of closing one eye gives us a truer impression of actual perception than the subjects' own evaluations. In short, if subjective evaluations are liable to reverse the depth order of stimuli from 'monocular < binocular' to 'monocular > binocular' then we have good reason to be sceptical of them.

But we still have to explain why the subjective evaluations of depth reversed the depth order? As we have already discussed, Todd and Norman suggest that the subjective evaluations had a strategic component. Specifically, they claim that their subjects had to consciously convert their perceived depth into physical depth by comparing it to the height and width of the displays. But the problem with this explanation is that this concern equally applies to both the monocular motion and the binocular disparity plus motion displays, so it doesn't explain why the translation of perceived depth into physical depth should have reversed the depth order between the monocular motion and binocular disparity plus motion displays.

Instead, I would argue that the subjects simply *misjudged* their own visual experience in the monocular motion display: they thought they saw more depth than they actually did. This is because their evaluation of own their perceptual experience is, like any cognitive process, open to being *biased* or *prejudiced* by the depth *depicted* by the monocular cues. As I explain in Chap. 3, we can only know how much depth a non-disparity display produces by viewing it synoptically (sending an identical image to both eyes) and then introducing various points with binocular disparity into the scene. And this is essentially what Todd and Norman got their subjects to do in reverse by closing one eye, with the depth from disparity throwing the comparative flatness of the pictorial cues into sharp contrast.

Indeed, Todd and Norman (2003) provide us with a valuable illustration of the dilemma facing the Cue Integration literature: either we (a) rely on subjective evaluations, in which case there is no guarantee that the subjects' evaluations of their own perceptual experience is accurate (and, following Ernst et al. 2000; Todd and Norman 2003, significant evidence that it is not), or (b) we attempt to make a direct comparison between the stimuli, which may work in some contexts (e.g.

Todd and Norman 2003), but which may raise apparent motion concerns others (e.g. Hillis et al. 2002).

3. Indirect Comparison: But what we are studying is not merely the mechanisms that underpin depth perception in the laboratory, but also the mechanisms that explain our perception of the real world. And in the real world, we rarely get a chance to view and evaluate objects in isolation; we have no choice but to gauge their geometry in the presence of other objects. So although there is evidence that *proximity* (Gogel 1956) and *framing* (Eby and Braunstein 1995) may influence our evaluations when objects are not viewed in isolation, these influences cannot be so pervasive that they render any such evaluation completely redundant. Which opens up the possibility of a *third* strategy:

Instead of asking subjects to (a) *directly compare* Stimulus A with Stimulus B, or (b) *subjectively evaluate* Stimulus A in isolation, and then Stimulus B in isolation, and then compare these evaluations, we might attempt to (c) *indirectly compare* Stimulus A to Stimulus B, by first comparing Stimulus A to Stimulus C and then Stimulus B to Stimulus C. Indeed, Stimulus C might well be a second object or visual element that persists *at the same time* as both Stimulus A and Stimulus B. Nor should *proximity* or *framing* overly concern us; to the extent these concerns are brought into play they ought to equally affect the comparison between Stimulus A and Stimulus C on the one hand and Stimulus B and Stimulus C on the other.

But what would be a suitable comparator? Well, given binocular disparity is a cue to depth *off the fronto-parallel plane*, it would be useful to have an object or cue that marked out the location of the fronto-parallel plane, against which we could judge the degree of stereopsis in the scene with a simple comparison. Ironically enough, just such a cue was introduced by Ernst et al. (2000) in Fig. 3. Notice the black crosses in the centre of the stimuli: these black crosses were not present in the original experiment, but were added to the published version of the stimuli to help readers cross-fuse. But in the context of our discussion these black crosses take on another role: since they lack disparity or perspective, they demarcate the fronto-parallel plane. Nor should the presence of these black crosses overly affect the cue-conflict stimuli themselves: both the crosses and the cue-conflict stimuli ought to be regarded as freestanding objects defined by their own perspective and disparity cues. Nor should the crosses affect one cue-conflict stimulus more than the other: if Cue Integration really does occur *prior* to form perception, it shouldn't

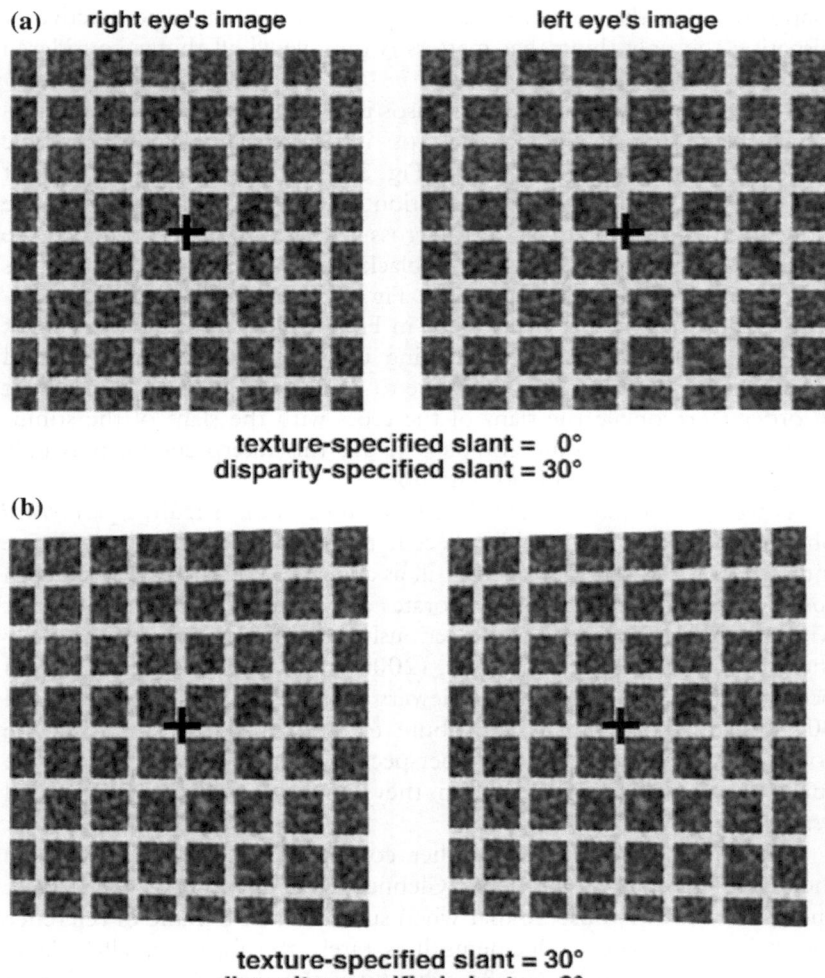

Fig. 3 Two examples of the cue-conflict stimuli from Ernst et al. (2000): **a** Texture specifies a slant of 0° whilst disparity specifies a slant of about 30° (at a viewing distance of 20 cm). **b** Texture specifies a slant of about 30° whilst disparity specifies a slant of 0°. From Ernst et al. (2000). Touch can change visual slant perception. Nature Neuroscience, 3(1), 69–73. © Nature Publishing Group

matter whether the perceived slant is primarily a product perspective or disparity; the only thing that matters is the overall all-things-considered determination.

And yet, even with all these provisos in place, this isn't how we experience the stimuli: in Fig. 3a, the cross is clearly slanted in stereoscopic space against the stimulus, whilst in Fig. 3b the cross is clearly flat against the stimulus. Indeed, this observation is only accentuated when we elongate the horizontal bars of the cross in (Fig. 4). So according to the quick and easy comparison that the black crosses and the horizontal bars afford us, the binocular disparity in Fig. 3a contributes positive stereoscopic slant, but the perspective cue in Fig. 3b does not. Nor do I think that this is an artefact of introducing the black crosses or horizontal bars: admittedly the fact that we have to read the stimulus as transparent in order to reconcile the slant of the cross with the slant of the stimulus might introduce some complexity, but this interpretation is readily adopted in Fig. 3a, so why not Fig. 3b?

As we have already observed, if Cue Integration is truly a *perceptual* phenomenon then we would expect it to be robust enough to survive interaction with other objects. But if, as appears to be the case, the depth specified by Cue Integration evaporates as soon as it comes into contact with another object, we have to seriously question whether it was really there in the first place. Ernst et al. (2000) suggest that Cue Integration occurs in Fig. 3b because 'most viewers perceive a slant between 0° and 30°, because both signals contribute to the perceived slant'. But we could just as easily imagine the perspective cues in Fig. 3b biasing the subjects' *evaluation* of the depth they perceive, but not their actual *perception*.

Finally, Fig. 3a illustrates another concern for Ernst et al.'s (2000) method of *subjective evaluation*: Glennerster et al. (2002), Glennerster and McKee (2004) observe that when subjects infer a frame of reference for their stereoscopic judgements it is rarely the fronto-parallel plane. This is demonstrated by Fig. 3a, where it is the cross that looks slanted relative to the stimulus, and not the other way round, further demonstrating just how poor our ability to evaluate stereoscopic depth really is.

To return to the question of pictorial cues biasing our *subjective evaluation* of stereoscopic depth, I would argue that the very same effect is evident not only when subjects (a) attribute depth to pictorial cues in the absence of binocular disparity (as in Ernst et al. 2000), but also (b) when subjects fail to attribute depth to small but otherwise discriminable

(a)   right eye's image   left eye's image

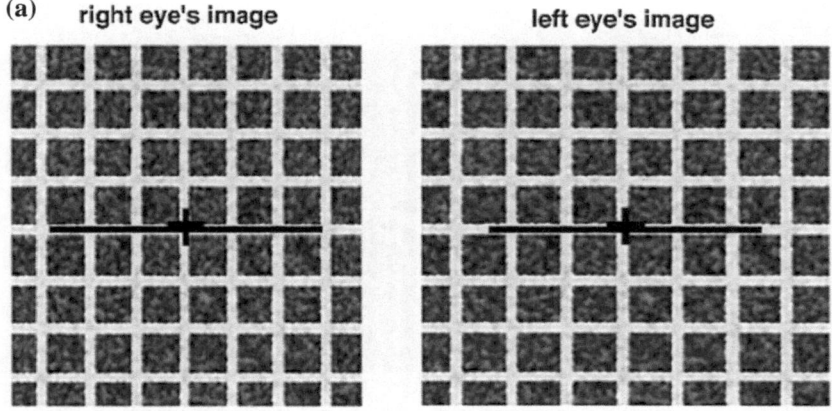

texture-specified slant =   0°
disparity-specified slant = 30°

(b)

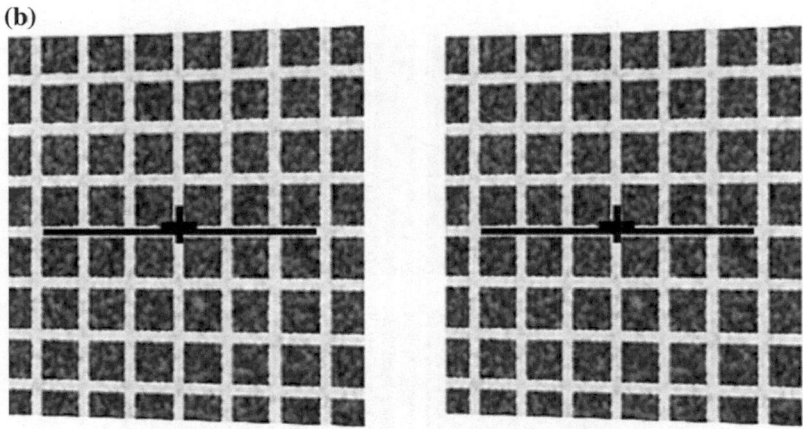

texture-specified slant = 30°
disparity-specified slant = 0°

Fig. 4   Fig. 3 with a horizontal bar added. Amended from Ernst et al. (2000). Touch can change visual slant perception. Nature Neuroscience, 3(1), 69–73. © Nature Publishing Group

(a)

(b)

(c)

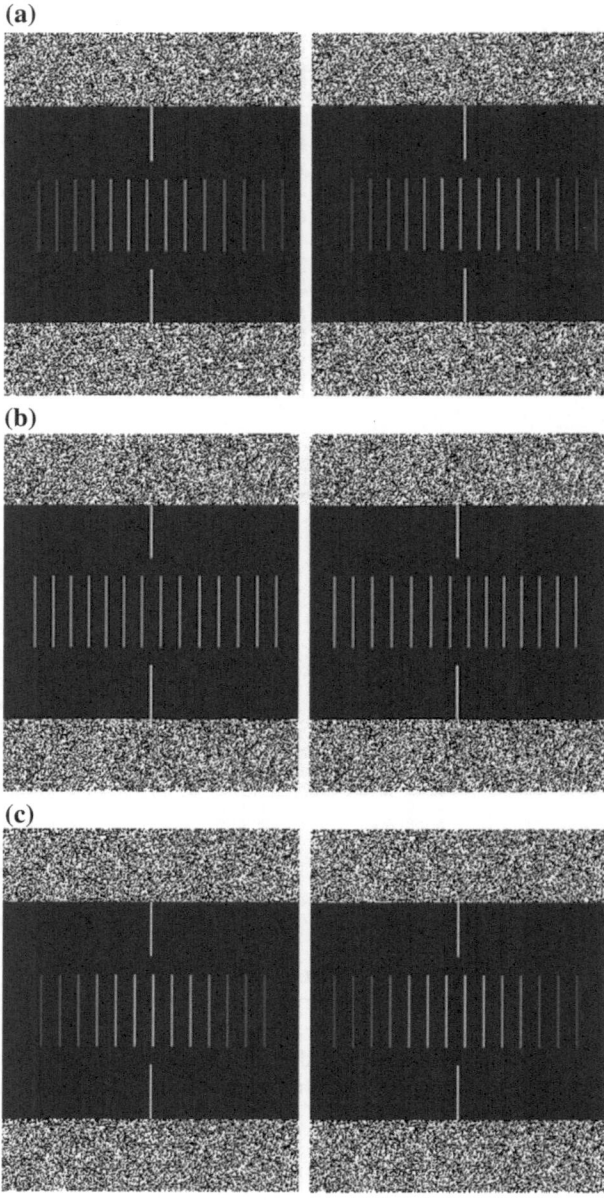

◆ **Fig. 5**   Stimuli from Likova and Tyler (2003): sparsely sampled Gaussian profiles defined by **a** luminance only, **b** disparity only, and **c** 'a combination of both cues at the level that produced a cancellation to flat plane under the experimental conditions'. From Likova and Tyler (2003). Peak localization of sparsely sampled luminance patterns is based on interpolated 3D object representations. Vision Research, 43, 2649–2657. © Elsevier

binocular disparities when conflicting pictorial cues are present, such as in Likova and Tyler (2003). Now there are two distinct questions in Likova and Tyler (2003) that I want to keep separate:

The first is how good we are at filling in on the basis of sparse information? For instance, if I give you the following sequence: 1, 2, 3, ..., 5, ..., 7, ..., you might immediately see that the missing numbers are 4, 6, and 8. And if I asked you to pick the highest value in the sequence you would point to the last ..., even though its value is not explicitly specified. Now according to Likova and Tyler (2003), it turns out that we are much better at inferring *3D form* from sparse information (e.g. Gregory's dalmatian: see Fig. 4 in Chap. 3) than we are at inferring *changes in surface luminance* from sparse information (e.g. Hume 1748's 'missing shade of blue' in a sequence of blue patches of increasing luminance); and Likova and Tyler demonstrate this fact with Fig. 5:

Consider the 14 vertical bars with varying luminance in the two (identical) images in Fig. 5a. We could interpret them as 14 individual vertical bars, or we could interpret them as part of a single continuous horizontal black-and-white surface. Even as a horizontal surface, Fig. 5a is open to two interpretations: a flat 2D surface with a change in luminance or a convex 3D surface cast in shadow.

First, Likova and Tyler found that in spite of the absence of foreshortening (see Hartle and Wilcox 2016), subjects automatically adopted the latter (3D) interpretation: the luminance profile evoked 'an unambiguous depth percept of the brighter bars appearing closer for all observers'. Second, Likova and Tyler found that subjects were actually quite good at interpolating the location of the surface bulge when it was interpreted as a bulge in 3D shape. Third, however, when this 3D interpretation was barred by a competing disparity profile, and subjects were left trying to interpolate the bulge as a change in the luminance of a 2D surface, they were unable to perform the task: 'Once the depth interpretation is nulled by the disparity signal, the luminance information does not support position discrimination at all'.

Now the second question that Likova and Tyler explore, and the one that concerns us, is their use of disparity to cancel or null the 3D interpretation of the luminance profile, leaving a surface with otherwise discriminable luminance and disparity cues looking flat. Likova and Tyler confirm this cancellation effect by testing the disparity at which the luminance profile looked flat, and found that this did not occur at zero disparity, but at a small negative disparity of between -0.3 and -0.4 arc min. For Likova and Tyler, this observation confirms the fact that 'the perceived depth from the luminance profile lies in the same qualitative dimension as the perceived depth from disparity cues (i.e. that it is a 'true' depth percept rather than just a cognitive inference of some kind)'. Now if by a 'cognitive inference' Likova and Tyler mean that subjects consciously subtract the depth from disparity from the depth from luminance (for instance, Likova and Tyler allude to the possibility that 'the luminance patterns might be interpreted as an object during localization'), then I quite agree. But as I have continually emphasised in this chapter, between the *perceptual bias* that Likova and Tyler argue for, and the *conscious decision-making* that Likova and Tyler reject, there is a third possibility, namely a *cognitive bias*: the idea that our evaluation of our own visual experience can be biased by the presence of confounding cues.

Such an interpretation is entirely consistent with Likova and Tyler's cancellation paradigm, especially since (a) the disparity that is cancelled is small (0.3–0.4 arc min), and (b) the determination of the null-point itself relies upon a cognitive judgement of comparative depth: as Likova and Tyler explain, the null-point is only approximate, and required subjects to judge whether the centre of the bulge appeared to be at the same depth as the bars on the far left and far right, even if some 'minor wrinkles' could be seen in the transition regions. But as I constantly emphasise, the judgement that something 'looks flat' or 'looks bulged' is exactly that: a *judgement*; and, as with all judgements, we need to ask what confidence we have in our ability to make these judgements without bias?

My interpretation would also be consistent with the results of Likova and Tyler's main study where the luminance profile introduced a small but consistent bias in favour of concave depth, but left depth from disparity otherwise intact (Fig. 6): As the curves fitted through the points in Fig. 6 (a) and (b) demonstrate, the effect of luminance is to shift the

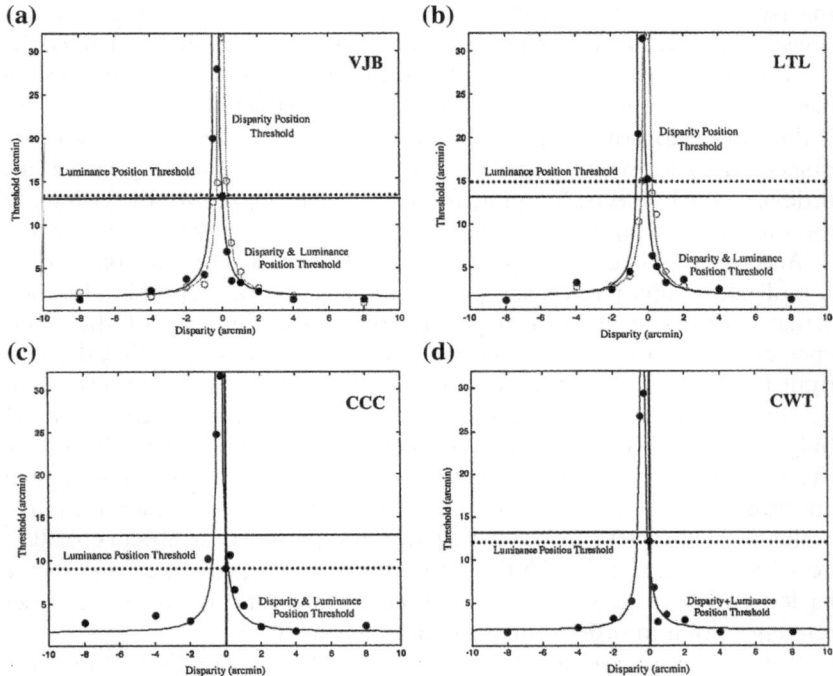

Fig. 6 The results of the position localisation task for the two principal observers in Likova and Tyler (2003): VJB and LTL, with key conditions verified with another two observers: CCC and CWT. The white circles are the thresholds for the disparity only condition, and the black circles are the thresholds for the disparity plus luminance condition. From Likova and Tyler (2003). Peak localization of sparsely sampled luminance patterns is based on interpolated 3D object representations. Vision Research, 43, 2649–2657. © Elsevier

whole psychometric function to the left by 0.3–0.4 arc min. As Likova and Tyler observe, this is the only change that luminance makes: 'all other aspects of the position task fell on the same curve with no change in parameter values'. So the results clearly demonstrate a fixed *bias* in extracting depth from disparity. But they don't determine whether this bias is *perceptual* or *cognitive*.

Third, there is one basis upon which we might try to determine whether the bias in Likova and Tyler is *perceptual* or *cognitive*. According to Likova and Tyler, the long-rage interpolation process that forms the basis of their study is the *only* means by which we can see everyday objects: they argue that since everyday objects are typically defined by local features separated by extended featureless regions the visual system has to engage in long-range interpolation to extract their 3D form. But if the integration of disparity and shading is really how we see everyday objects, then we would expect Likova and Tyler's stimuli to behave like ordinary visual objects: bringing an independent object into the vicinity of the surfaces shouldn't turn a *convex* surface *flat* and a *flat* surface *concave*.

And yet this is exactly what appears to happen when we place their stimuli alongside a reference point or alongside one another. In their actual experiment, Likova and Tyler offset the disparity of the reference point as a way of ensuring that subjects were interpolating the left-right location of the bulge. By contrast, the question that concerns us is whether luminance nulls depth from disparity? I.e. whether there is a 3D bulge in the first place. And in this context, there is no harm in setting the reference point at zero disparity as a test of true flatness: rather than engaging in what is, by Likova and Tyler's own admission, a long-range evaluative judgement in order to judge the presence or absence of flatness (by comparing the depth of the central bar against the bars on the far left and far right), what harm could it do to afford subjects a closer reference point in order to make their determinations?

But, as soon as we do, the *concavity* specified in Fig. 5c by binocular disparity (0.4 arc min) becomes fully apparent (Fig. 7).

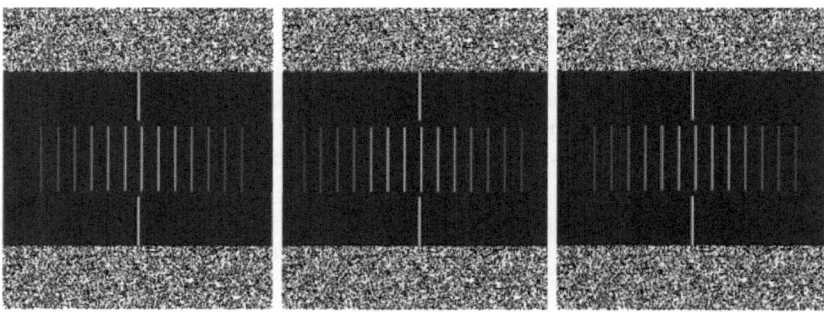

Fig. 7   Fig. 5c with disparity removed from the reference point, and the reference point brought closer to the stimulus. Amended from Likova and Tyler (2003). Peak localization of sparsely sampled luminance patterns is based on interpolated 3D object representations. Vision Research, 43, 2649–2657. © Elsevier

Fig. 8   Fig. 5a (top) added to Fig. 5c (bottom) with disparity removed from the reference point, and the reference and the stimuli brought closer together. Amended from Likova and Tyler (2003). Peak localization of sparsely sampled luminance patterns is based on interpolated 3D object representations. Vision Research, 43, 2649–2657. © Elsevier

Furthermore, the *flatness* specified in Fig. 5a by the absence of binocular disparity also becomes immediately apparent when it is added (upper stimulus) (Fig. 8).

None of the stimuli are occluded, so each ought to persist in its own depth defined by its own depth cues. So why, once we afford subjects a more accurate reference point by which to judge the *presence* or *absence* of stereoscopic depth, does depth from shading appear to evaporate?

3   ILLUSIONS

Having altered the cue-conflict stimuli in Ernst et al. (2000) and Likova and Tyler (2003) in order to better understand their *perceived* rather than merely *conceived* depth, we might wonder how a similar technique would affect the real-world cue conflicts encountered in Reverspectives (Fig. 4 in Chap. 1) and the hollow-face illusion? For instance, if we add horizontal and/or vertical bars to these illusions, what happens? Do the bars cut through the illusory depth? Or do they break the illusion altogether? (Figs. 9, 10)

In fact, neither occurs. Instead, the illusion persists *but the inverted depth is located behind the horizontal and vertical bars*: the inverted depth does not protrude beyond the bars even though as an inverted depth percept it ought to. It is as if the *inverted percept* and *stereoscopic space* are simply talking past one another. And I would argue that the only way to make sense of disconnect is to recognise that the inverted depth percept is, in fact, a *false judgement* that we apply to a *veridical percept* of the hollow face or Reverspective. In which case, the hollow Face and Reverspectives are better thought of as *delusions* rather than *illusions*: *misinterpretations* of what we see, rather than *false percepts*.

This observation opens up a whole new experimental strategy in trying to understand Reverspectives and hollow-face illusion. Indeed, we can place objects not just *in front*, but also at various points *inside*, the Reverspective and the hollow face, in space that ought not to exist according to the illusory percept. Does this destroy the illusion? Again, it doesn't appear to: swaying back and forth, we still get the 'illusory percept', in spite of the fact that we are also aware that we are viewing a hollow filled with objects. Indeed, we might distribute points within the hollow space, or place an object with an identifiable slant, in order to see how, if at all, our ordinal depth judgements are affected? For instance, if we place a pen at a slant in the hollow of the Reverspective, my experience is that we continue to see it as slanted in the right direction even though the 'illusory percept' persists (Fig. 11).

This effect would have to be confirmed experimentally, and perhaps the best test would be a giant (1.5–2 m) Reverspective or hollow face: on the one hand, the binocular disparity of the structure will be reduced, so we should be able to get closer whilst still maintaining the *global* inverted depth illusion, but on the other hand, if we use markers distributed in space the depth separation between these points will be increased, thereby accentuating the *ordinal* depth we have to judge.

Admittedly, we might see the pen move in Fig. 11, so do we at least get an illusory percept of motion, if not an illusory percept of depth? I would resist this conclusion not only because illusory depth and illusory motion appear to be two sides of the same coin (if one is a judgement,

Fig. 9   Reverspective with horizontal bar attached. Reverspective courtesy of http://www.offthewallartprints.com. © Off The Wall Art Prints

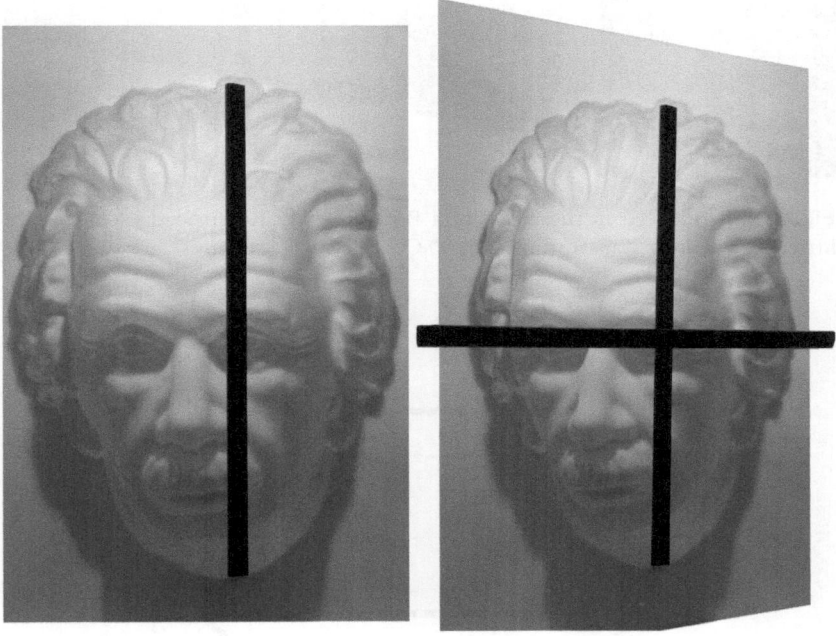

Fig. 10   Hollow-face illusion with vertical and horizontal bars attached. Hollow-face illusion courtesy of http://www.grand-illusions.com. © Grand Illusions

Fig. 11 Pen placed in the recess of a Reverspective. Reverspective courtesy of http://www.offthewallartprints.com. © Off The Wall Art Prints

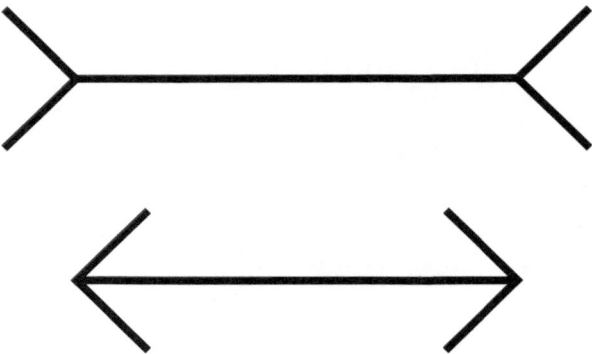

Fig. 12 Müller-Lyer illusion

then it would appear to follow that the other must be as well), but also because I am unconvinced that we truly *see* motion, as opposed to merely *judge* it, in the first place. This is not just a thesis about *illusory* motion, but motion altogether, and will have to be developed and defended in later work. But to give an illustrative example, consider the motion in the rotating dancer illusion: The literature tends to focus on the fact that it is bistable, i.e. that it is liable to switch from clockwise to counterclockwise. But the deeper point is this: we have a 3D rotation in a 2D image; we don't see the dancer as moving *laterally*, but as *rotating*. But we also don't get an impression of the dancer's leg extending beyond the computer screen: there is no stereopsis, it remains a 2D impression. So we paradoxically 'see' motion in a dimension that we do not literally see. A more satisfactory explanation, and one that coheres with my model of pictorial images (see Chap. 3), is that the motion of the dancer is also a post-perceptual unconscious inference rather than something that we see.

Returning to the question of illusory depth, another way of testing my hypothesis is to view the Reverspective monocularly whilst swaying back and forth. Certainly, I can get within 10 cm of the cardboard version I have before the illusion breaks. But at that distance, something weird begins to happen: as I focus on a recess in the Reverspective (that is painted as a protruding building) I get the illusory percept, but I also get the impression of the two physical peaks (that should be the furthest away according to the illusory percept) looming in and out of my vision as I sway back and forth. This is new observation so far as the literature is concerned.

One commentator has suggested to me that this might be an indication that at close distances we experience a mixed percept where *both* the real peaks of the Reverspective *and* the illusory peaks (or real troughs) of the Reverspective are now seen as peaks; leaving the perceived troughs midway down the inverted pyramid, halfway between the real peak and the real trough. But I don't think that this is the correct interpretation. After all, we can place an object (like a pencil) where this illusory trough ought to be and it becomes immediately apparent that the pencil is midway in depth between the real peak and the real trough; just as we would expect from a veridical percept.

But how are we to make sense of the suggestion that it is our *cognition* rather than our *perception* that is being misled by Reverspectives and the hollow-face illusion? Well, consider the question that Wittgenstein posed to Anscombe (recounted in Anscombe 1959):

> He once greeted me with the question: 'Why do people say that it was nat-
> ural to think that the sun went round the earth rather than that the earth
> turned on its axis?' I replied: 'I suppose, because it looked as if the sun
> went round the earth.' 'Well,' he asked, 'what would it have looked like if
> it had *looked* as if the earth turned on its axis?

The point being that we don't persist in thinking there is an *illusion*
of the sun going round the earth. Instead, we have come to recognise
that the same percept can have two interpretations, one of which may
seem more natural but ultimately turns out to be false. Similarly, recall
Gregory's (1970) classic demonstration of the hollow-face illusion of a
mask rotating on a stick. I would argue the hollow mask turning from
right-to-left looks exactly as it would if it *looked* like a hollow mask turning from right-to-left; the only difference is that we *judge* it to be a
protruding mask turning from left-to-right. We might come to this conclusion in two steps: First, I would argue that our visual experience of
a 2D video of the hollow-face illusion is entirely consistent with both
interpretations, and that we merely *judge* (rather than *perceive*) the mask
to have illusory depth and motion. Second, in the case of a real mask,
this principle is taken one step further: we judge the mask to have an
illusory depth and motion in spite of the fact that, to the extent that stereopsis is present, it points us towards a veridical interpretation. This is
just another instance of our *cognition* of a scene or object outstripping
our *perception* of it (which, as we have already discussed, may be artefact
of our evolutionary need to *interpret* the visual scene as invariant, even
though our *perception* of stereopsis falls-off with distance).

But what about the case where we sway back and forth in front of
the object: when we have a veridical percept the Reverspective or hollow face appears to remain fixed, but when we have an illusory inverted
percept, the Reverspective or hollow face appears to follow us around the
room. But again, I would argue that whether something appears fixed
or appears to move is not *seen*, but a *post-perceptual inference* applied to
what we see. Take the classic case of illusory self-motion: a neighbour-
ing train pulls away from the station, and you mistakenly believe that it
is *your* train that is in motion: the veridical interpretation is an equally
permissible interpretation of what we see; there is nothing *in our visual
experience* that identifies the illusory interpretation over the veridical one.

In conclusion, Gregory (1997) observed that: 'To maintain that
perception is direct, without need of inference or knowledge, Gibson

generally denied the phenomena of illusion'. But in this section I have argued that the denial of *illusions* (as opposed to *delusions*) is not driven by ideological commitments, but by our actual experience: how can we be experiencing an illusory percept if, to the extent we are able to measure the ordinal depth of various points on the Reverspective or the hollow face, they all turn out to be veridical? This account is quite a departure from the contemporary literature where even a purely cognitive explanation of the Müller-Lyer illusion (Fig. 12) is not only assumed to be false, but *obviously* or *self-evidently* false: see Morgan et al. (2012), Witt et al. (2015), and in the Philosophical context Phillips (2016). These articles argue that the tails of the lines in the Müller-Lyer illusion bias our *perception* of their length, whilst I would argue that they merely bias our *post-perceptual evaluation* of their length.

Gregory's work on illusions was influenced by his encounter with Sidney Bradford, a man born blind whose sight was restored following an operation (see Gregory and Wallace 1963; Gregory 2004). Faced with the apparent ineffectiveness of illusions on Sidney Bradford, Gregory concluded that many illusions were acquired over time and must be *cognitive* in nature. Gregory could have drawn one of two implications from this conclusion: either (a) that these illusions were *merely cognitive* rather than perceptual, or (b) that since these illusions were perceptual, perception itself must be cognitive. We all know that Gregory chose the latter interpretation, but is the former really that unsustainable?

4   ATYPICAL RESPONSES TO CUE INTEGRATION

But Sidney Bradford's is not the only atypical response to cue-conflict stimuli:

1. Binocular Depth Inversion Illusion (BDII): If we take a stereogram of a human face and reverse the images, the resulting face will *not* ordinarily be seen as a hollow face by normal observers. This failure of pseudoscopy is often interpreted as pictorial cues vetoing an *unlikely* percept from binocular disparity. By contrast, I would be reluctant to embrace this interpretation before first confirming that the inverted binocular disparity specified a *coherent* depth percept (for instance, simply switching stereo photographs of a real human face—which is the usual stimulus in this context—is liable to introduce discontinuities whenever there is an overhang, and so any 'vetoing' in this context could simply reflect the

fact that no coherent 3D surface can be constructed out of the disparity information). Similarly, there is a question of the extent to which we are merely tracking a *failure* of pseudoscopy (with the eventual percept appearing flat, or at best merely pictorial), rather than a *positive* inversion of depth (as we experience with the hollow-face illusion).

Nonetheless, it has been demonstrated that subjects have a *more accurate* (and *less illusory*) percept of BDII stimuli in the context of (a) schizophrenia (Emrich 1989; Schneider et al. 1996a; Schneider et al. 2002; Koethe et al. 2006, 2009; Dima et al. 2009; Keane et al. 2013, 2016; Gupta et al. 2016), (b) cannabis (Emrich et al. 1991; Leweke et al. 2000; Semple et al. 2003; Koethe et al. 2006), (c) alcohol (Schneider et al. 1998), (d) alcohol withdrawal (Schneider et al. 1996b, 1998), (e) anxiety (Passie et al. 2013), and even (f) sleep deprivation (Schneider et al. 1996a; Sternemann et al. 1997). (Keane et al. 2013, 2016; Gupta et al. 2016 utilise a hollow mask, with which I have no methodological complaint). By contrast, no statistically significant effect was found in the presence of (g) bipolar (Koethe et al. 2009, 2016), (h) dementia (Koethe et al. 2009), (i) depression (Schneider et al. 2002; Koethe et al. 2009), or (j) ketamine (Passie et al. 2003).

But these results appear to pose the following question: just how often in our daily lives would it be detrimental to see the world *more* accurately? For instance, Keane et al.'s (2016) central contention is that those with schizophrenia are able to see the world 'more clearly through psychosis'. But it is difficult to see why this perceptual advantage (specifically, being able to 'more accurately perceive object depth structure') should be problematic? By contrast, being unable to attribute proper context or meaning to whatever we see (my *cognitive* rather than *perceptual* explanation for the failure of depth inversion in this context) would be problematic; leaving those subject to schizophrenia unable to rely on past experience and reliant on ad-hoc rationalisations as they try to make sense of what they see.

2. Child Development: As the case of Sidney Bradford demonstrates, the ability to interpret pictorial cues has to be acquired by experience. For children, this appears to occur around 5–7 months; as Arterberry (2008) explains, by this age children are typically able to rely upon shading, linear perspective, occlusion, texture gradients, familiar size, and surface contours to guide their reaching responses. Depth from disparity also emerges during early child development (around 3–5 months, according to Fox et al. 1980; Held et al. 1980; and Birch et al. 1982).

But as Hong and Park (2008) demonstrate, depth from disparity is relatively coarse for the first 3–4 years (0.68 arc min), and only matures to adult levels of stereoacuity (0.23 arc min) at year 5. Consequently, the early years of child development represent a prime opportunity for Cue Integration to compensate for poor stereoacuity using pictorial cues.

And yet, as Nardini et al. (2010) demonstrate, what is startling is just how late Cue Integration emerges in children. Nardini et al. estimate that Cue Integration of texture and disparity only begins to emerge at around 12 years of age. Certainly, the 6 year olds that they tested did not experience Cue Integration when presented with the cue-conflict stimuli from Hillis et al. (2002). Indeed, the children were able to outperform adults when the sources of information were in conflict, due to an absence of *mandatory fusion*. (And it is worth noting that even for the adults *mandatory fusion* was not complete). So the question is why adults lose access to the single cues relative to children? Given that this loss occurs so late, it doesn't seem plausible that at 12 years of age the processes that govern stereopsis suddenly switches from binocular disparity to Cue Integration. More plausible is the suggestion that experience biases our ability to accurately evaluate the various components of our perception, which would explain the adults' underperformance.

3. Autistic Adolescents: Building on their work with 6 year olds, Bedford et al. (2016) applied the same test to autistic adolescents (12–15 year olds), and found that adolescents with autism integrate cues when it is to their advantage (for instance, two sub-threshold changes in the same direction), but not when it would lead to a reduction in performance (e.g. two opposing changes could be discriminated, rather than cancelling each other out). Bedford et al. suggest that this is a new pattern of behaviour, which they term 'selective fusion', but we've seen this pattern before:

4. Cross-Modal (Vision and Touch): Whilst Hillis et al. (2002) reported mandatory fusion in the context of visual depth cues such as texture and disparity, they came to the apparently paradoxical conclusion that we both *have* and *do not have* perceptual fusion in the cross-modal context of vision and touch:

> We also have evidence for *a single, fused percept* for shape information from haptics and vision, *but* in this intermodal case *information from single-cue estimates is not lost*. (emphasis added)

What is going on here? Well, as with Bedford et al., Hillis et al. distinguish between cases where (a) cue combination is likely to lead to an *improved* performance (e.g. two sub-threshold changes in the *same* direction), and (b) cases where cue combination is likely to lead to a *worse* performance (e.g. two above-threshold changes in *opposite* directions), and suggest that subjects only experience Cue Integration when it is to their benefit: i.e. in (a) but not (b).

So, in at least two contexts (autistic adolescents and cross-modal perception), *mandatory fusion* begins to look somewhat less than mandatory. But remember that mandatory fusion (the idea that there are *costs* as well as *benefits* to Cue Integration) was introduced by Hillis et al. (2002) as a means of convincing us that Cue Integration must be operating at the level of *perception* rather than *cognition*. By contrast, the absence of mandatory fusion in (a) autistic adolescents and (b) cross-modal perception appears to suggest that when information is integrated in these contexts it is by virtue of an *integrated judgement* rather than an *integrated percept*. After all, you cannot choose to *see* something as a single, fused percept when it is to your benefit, but as disparate sources of information when it is not. But, and this is the important point, if an *integrated judgement* can explain Cue Integration in the context of (a) autistic adolescents, and (b) cross-modal perception, what makes us so confident that it doesn't explain Cue Integration more broadly?

5. Are we all Atypical? As Oruç et al. (2003) observe, 'large individual differences are the rule in depth perception studies'. But these individual differences are often passed over in the literature without comment: e.g. Hillis et al. (2002), Knill (2007). By contrast, two papers that did make these individual differences a central component were Oruç et al. (2003) and Zalevski et al. (2007). Indeed, the individual differences in Zalevski et al. were so pervasive that: 'No definite conclusion could be drawn ... because of the large variability associated with the estimated weights'. And so, in an ironic twist, the individual differences themselves became the major finding of that study:

> The large individual differences we found in cue-combination studies suggest that human observers differ in their cue-weighting strategies and it may be that there is no single model to account for all behaviour, especially when cues to depth are few and in conflict.

Similarly, Oruç et al. (2003) found that although the performance of 6 out of their 8 subjects was consistent with optimal Cue Integration, 2 out of their 8 subjects were positively inconsistent with it. Indeed, these 2 subjects did not even appear to benefit from having two cues rather than one. And it was in light of this, and similar findings in the literature, that Oruç et al. concluded that 'individual differences abound in depth perception studies (such as the widely ranging cue reliabilities across subjects found by Hillis et al. 2002...)'. What this observation demonstrates is that even the most fervent advocates of Cue Integration (such as Landy and Maloney) recognise that individual differences pose a real concern for their account.

The only alternative is to argue that these individual differences are predicted by Cue Integration itself: for instance, Marty Banks has suggested that Ernst and Banks (2002), Hillis et al. (2004), and Girshick and Banks (2009) demonstrate that different subjects attach different weights to cues based upon the different reliabilities of these cues *for them*. The implication being that although the subjects behave differently, they were all, in fact, exhibiting optimal Cue Integration *given the different reliabilities of each individual cue for each individual subject*. But this raises the question as to why the reliability of each individual cue should vary so drastically between individual subjects (in the way that would be required to account for the results in Oruç et al. 2003; Zalevski et al. 2007)?

But my primary concern is not to pose problems for *optimal* Cue Integration, but to question the level at which Cue Integration occurs in the first place? If Cue Integration really does occur at the level of *perception* then, in light of these pervasive individual differences, we necessarily commit ourselves to the suggestion that even 'normal' subjects perceive substantially different slants and angles from one another, with the same scene presented to each observer with an idiosyncratic geometry. By contrast, if Cue Integration merely occurs at the level of *cognition*, then all we have to maintain is that subjects are liable to have idiosyncratic *interpretations* of the very same perceived geometry. The former really is quite a radical conclusion to have to come to in order to accommodate the pervasive individual differences in the literature. By contrast, the latter is exactly what we would expect; namely, that subjects are liable to *interpret* what they see in a variety of different ways.

REFERENCES

Albertazzi, L., van Tonder, G. J., & Vishwanath, D. (2010). *Perception beyond inference: The informational content of visual processes.* Cambridge, MA: MIT Press.

Ames, A., Jr. (1951). Visual perception and the rotating trapezoidal window. *Psychological Monographs, 65*(7), 324.

Ames, A., Jr. (1955). *An interpretative manual: The nature of our perceptions, prehensions, and behavior.* For the demonstrations in the Psychology Research Center, Princeton University. Princeton, NJ: Princeton University Press.

Anscombe, G. E. M. (1959). *An introduction to Wittgenstein's Tractatus.* London: Hutchinson.

Arterberry, M. E. (2008). Infants' sensitivity to the depth cue of height-in-the-picture-plane. *Infancy, 13*(5), 544–555.

Bedford, R., Pellicano, E., Mareschal, D., & Nardini, M. (2016). Flexible integration of visual cues in adolescents with autism spectrum disorder. *Autism Research, 9*(2), 272–281.

Birch, E. E., Gwiazda, J., & Held, R. (1982). Stereoacuity development for crossed and uncrossed disparities in human infants. *Vision Research, 22*(5), 507–513.

Cavanagh, P. (2011). Visual cognition. *Vision Research, 51*(13), 1538–1551.

Chen, C. C., & Tyler, C. W. (2015). Shading beats binocular disparity in depth from luminance gradients: Evidence against a maximum likelihood principle for cue combination. *PLoS One, 10*(8), e0132658.

Dima, D., Roiserc, J. P., Dietricha, D. E., Bonnemanna, C., Lanfermannd, H., Emricha, H. M., et al. (2009). Understanding why patients with schizophrenia do not perceive the hollow-mask illusion using dynamic causal modelling. *NeuroImage, 46*(4), 1180–1186.

Domini, F., & Caudek, C. (2011). Combining image signals before three-dimensional reconstruction: The intrinsic constraint model of cue integration. In Trommershäuser, Körding, & Landy (Eds.), *Sensory cue integration.* Oxford: Oxford University Press.

Doorschot, P. C., Kappers, A. M., & Koenderink, J. J. (2001). The combined influence of binocular disparity and shading on pictorial shape. *Perception and Psychophysics, 63*, 1038–1047.

Eby, D. W., & Braunstein, M. L. (1995). The perceptual flattening of three-dimensional scenes enclosed by a frame. *Perception, 24*(9), 981–993.

Emrich, H. M. (1989). A three-component-system-hypothesis of psychosis. Impairment of binocular depth inversion as an indicator of functional dysequilibrium. *British Journal of Psychiatry, 155*(S5), 37–39.

Emrich, H. M., Weber, M. M., Wendl, A., Zihl, J., von Meyer, L., & Hanisch, W. (1991). Reduced binocular depth inversion as an indicator of cannabis-induced censorship impairment. *Pharmacology, Biochemistry and Behavior, 40*(3), 689–690.

Ernst, M. O., & Banks, M. S. (2002). Humans integrate visual and haptic information in a statistically optimal fashion. *Nature, 415*(6870), 429–433.

Ernst, M. O., & Bülthoff, H. H. (2004). Merging the senses into a robust percept. *Trends in Cognitive Sciences, 8*(4), 162–169.

Ernst, M. O., Banks, M. S., & Bülthoff, H. H. (2000). Touch can change visual slant perception. *Nature Neuroscience, 3*(1), 69–73.

Fox, R., Aslin, R. N., Shea, S. L., & Dumais, S. T. (1980). Stereopsis in human infants. *Science, 207*(4428), 323–324.

Gepshtein, S., Burge, J., Ernst, M. O., & Banks, M. S. (2005). The combination of vision and touch depends on spatial proximity. *Journal of Vision, 5*(11), 1013–1023.

Girshick, A. R., & Banks, M. S. (2009). Probabilistic combination of slant information: Weighted averaging and robustness as optimal percepts. *Journal of Vision, 9*(9), 8.

Glennerster, A., & McKee, S. P. (2004). Sensitivity to depth relief on slanted surfaces. *Journal of Vision, 4*, 378–387.

Glennerster, A., McKee, S. P., & Birch, M. D. (2002). Evidence for surface-based processing of binocular disparity. *Current Biology, 12*, 825–828.

Gogel, W. C. (1956). The tendency to see objects as equidistant and its inverse relation to lateral separation. *Psychological Monographs: General and Applied, 70*(4), 1–17.

Gregory, R. L. (1970). *The intelligent eye.* London: Weidenfeld & Nicolson.

Gregory, R. L. (1997). Knowledge in perception and illusion. *Philosophical Transactions of the Royal Society B, 352*, 1121–1128.

Gregory, R. L. (2004). The blind leading the sighted. *Nature, 430*, 1.

Gregory, R. L., & Wallace, J. G. (1963). *Recovery from early blindness: A case study.* Experimental Psychology Society Monograph No. 2. Cambridge: Heffer.

Gupta, T., et al. (2016). Disruptions in neural connectivity associated with reduced susceptibility to a depth inversion illusion in youth at ultra high risk for psychosis. *NeuroImage, 12*, 681–690.

Hartle, B., & Wilcox, L. M. (2016). Depth magnitude from stereopsis: Assessment techniques and the role of experience. *Vision Research, 125*, 64–75.

Held, R., Birch, E., & Gwiazda, J. (1980). Stereoacuity of human infants. *Proceedings of the National Academy of Sciences, 77*(9), 5572–5574.

Held, R. T., Cooper, E. A., & Banks, M. S. (2012a). Blur and Disparity Are Complementary Cues to Depth. *Current Biology, 22*(5), 426–431.

Held, R. T., Cooper, E. A., & Banks, M. S. (2012b). *Response to Vishwanath.* Originally published alongside Vishwanath (2012a) online, currently unavailable.

Hillis, J. M., Ernst, M. O., Banks, M. S., & Landy, M. S. (2002). Combining sensory information: Mandatory fusion within, but not between, senses. *Science, 298,* 1627–1630.

Hillis, J. M., Watt, S. J., Landy, M. S., & Banks, M. S. (2004). Slant from texture and disparity cues: Optimal cue combination. *Journal of Vision, 4*(12), 967–992.

Hong, S. W., & Park, S. C. (2008). Development of distant stereoacuity in visually normal children as measured by the Frisby-Davis distance stereotest. *British Journal of Ophthalmology, 92,* 1186–1189.

Hume, D. (1748). *An Enquiry Concerning Human Understanding.* London: A. Millar.

Keane, B. P., Silverstein, S. M., Wang, Y., & Papathomas, T. V. (2013). Reduced depth inversion illusions in schizophrenia are state-specific and occur for multiple object types and viewing conditions. *Journal of Abnormal Psychology, 122*(2), 506–512.

Keane, B. P., Silverstein, S. M., Wang, Y., Roché, M. W., & Papathomas, T. V. (2016). Seeing more clearly through psychosis: Depth inversion illusions are normal in bipolar disorder but reduced in schizophrenia. *Schizophrenia Research, 176*(2), 485–492.

Knill, D. C. (2007). Learning Bayesian priors for depth perception. *Journal of Vision, 7*(8), 13.

Knill, D. C., & Richards, W. (1996). *Perception as Bayesian inference.* Cambridge: Cambridge University Press.

Koenderink, J. J. (2010). Vision and information. In Albertazzi, van Tonder, & Vishwanath (Eds.), *Perception beyond inference: The informational content of visual processes.* Cambridge, MA: MIT Press.

Koethe, D., et al. (2006). Disturbances of visual information processing in early states of psychosis and experimental delta-9-tetrahydrocannabinol altered states of consciousness. *Schizophrenia Research, 88*(1–3), 142–150.

Koethe, D., et al. (2009). Binocular depth inversion as a paradigm of reduced visual information processing in prodromal state, antipsychotic-naïve and treated schizophrenia. *European Archives of Psychiatry and Clinical Neuroscience, 259,* 195.

Landy, M. S., Maloney, L. T., & Young, M. J. (1991). Psychophysical estimation of the human depth combination rule. In P. S. Schenker (Ed.), *Sensor fusion III: 3-D perception and decognition, Proceedings of the SPIE, 1383* (pp. 247–254).

Landy, M. S., Maloney, L. T., Johnston, E. B., & Young, M. (1995). Measurement and modeling of depth cue combination: In defense of weak fusion. *Vision Research, 35*, 389–412.

Landy, M., Banks, M., & Knill, D. (2011). Ideal-observer models of cue integration. In Trommershäuser, Körding, & Landy (Eds.), *Sensory cue integration*. Oxford: Oxford University Press.

Leweke, F. M., Schneider, U., Radwana, M., Schmidt, E., & Emrich, H. M. (2000). Different effects of nabilone and cannabidiol on binocular depth inversion in man. *Pharmacology, Biochemistry and Behavior, 66*(1), 175–181.

Likova, L. T., & Tyler, C. W. (2003). Peak localization of sparsely sampled luminance patterns is based on interpolated 3D object representations. *Vision Research, 43*, 2649–2657.

Maloney, L. T., & Landy, M. S. (1989). A statistical framework for robust fusion of depth information. In W. A. Pearlman (Ed.), *Visual communications and image processing IV, Proceedings of the SPIE, 1191* (pp. 1154–1163).

Morgan, M., Dillenburger, B., Raphael, S., & Solomon, J. A. (2012). Observers can voluntarily shift their psychometric functions without losing sensitivity. *Attention, Perception, & Psychophysics, 74*, 185–193.

Nardini, M., Bedford, R., & Mareschal, D. (2010). Fusion of visual cues is not mandatory in children. *Proceedings of the National Academy of Sciences, 107*(39), 17041–17046.

Ogle, K. (1959). The theory of stereoscopic vision. In S. Koch (Ed.), *Psychology: A study of a science (vol. I). Sensory, perceptual and physiological formulations* (pp. 362–394). New York: McGraw Hill.

Oruç, I., Maloney, L. T., & Landy, M. S. (2003). Weighted linear cue combination with possibly correlated error. *Vision Research, 43*, 2451–2468.

Passie, T., Karst, M., Borsutzky, M., Wiese, B., Emrich, H. M., & Schneider, U. (2003). Effects of different subanaesthetic doses of (S)-ketamine on psychopathology and binocular depth inversion in man. *Journal of Psychopharmacology, 17*(1), 51–56.

Passie, T., Schneider, U., Borsutzky, M., Breyer, R., Emrich, H. M., Bandelow, B., et al. (2013). Impaired perceptual processing and conceptual cognition in patients with anxiety disorders: A pilot study with the binocular depth inversion paradigm. *Psychology, Health, & Medicine, 18*(3), 363–374.

Phillips, I. (2016). Naive realism and the science of (some) illusions. *Philosophical Topics, 44*(2), 353–380.

Scarfe, P., & Hibbard, P. B. (2011). Statistically optimal integration of biased sensory estimates. *Journal of Vision, 11*(7), 1–17.

Schneider, U., Leweke, F. M., Sternemann, U., Emrich, H. M., & Weber, M. W. (1996a). Visual 3D illusion: A systems-theoretical approach to psychosis. *European Archives of Psychiatry and Clinical Neuroscience, 246*(5), 256–260.

Schneider, U., Leweke, F. M., Niemcyzk, W., Sternemann, U., Bevilacqua, M., & Emrich, H. M. (1996b). Impaired binocular depth inversion in patients with alcohol withdrawal. *Journal of Psychiatric Research, 30*(6), 469–474.

Schneider, U., Dietrich, D. E., Sternemann, U., Seeland, I., Gielsdorf, D., Huber, T. J., et al. (1998). Reduced binocular depth inversion in patients with alcoholism. *Alcohol and Alcoholism, 33*(2), 168–172.

Schneider, U., Borsutzky, M., Seifert, J., Leweke, F. M., Huber, T. J., Rollnik, J. D., et al. (2002). Reduced binocular depth inversion in schizophrenic patients. *Schizophrenia Research, 53*(1–2), 101–108.

Semple, D. M., Ramsden, F., & McIntosh, A. M. (2003). Reduced binocular depth inversion in regular cannabis users. *Pharmacology, Biochemistry and Behavior, 75*(4), 789–793.

Sternemann, U., Schneider, U., Leweke, F. M., Bevilacqua, C. M., Dietrich, D. E., & Emrich, H. M. (1997). Propsychotic change of binocular depth inversion by sleep deprivation. *Nervenarzt, 68*(7), 593–596.

Todd, J. T., & Norman, J. F. (2003). The visual perception of 3-D shape from multiple cues: Are observers capable of perceiving metric structure? *Perception & Psychophysics, 65*, 31–47.

Tyler, C. W. (2004). Theory of texture discrimination of based on higher-order perturbations in individual texture samples. *Vision Research, 44*(18), 2179–2186.

Vishwanath, D. (2005). The epistemological status of vision science and its implications for design. *Axiomathes, 15*(3), 399–486.

Vishwanath, D., & Domini, F. (2013). Pictorial depth is not statistically optimal. *Journal of Vision, 13*(9), 613.

Vishwanath, D., & Hibbard, P. B. (2013). Seeing in 3D with just one eye: Stereopsis without binocular vision. *Psychological Science, 24*(9), 1673–1685.

Witt, J. K., Taylor, J. E. T., Sugovic, M., & Wixted, J. T. (2015). Signal detection measures cannot distinguish perceptual biases from response biases. *Perception, 44*, 289–300.

Young, M. J., Landy, M. S., & Maloney, L. T. (1993). A perturbation analysis of depth perception from combinations of texture and motion cues. *Vision Research, 3*(18), 2685–2696.

Zalevski, A. M., Henning, G. B., & Hill, N. J. (2007). Cue combination and the effect of horizontal disparity and perspective on stereoacuity. *Spatial Vision, 20*(1–2), 107–138.

CHAPTER 3

Stereopsis in the Absence of Binocular Disparity

Abstract The absence of binocular disparity can make the world seem substantially flatter (for instance, when we close one eye) or even, I would argue, completely flat (for instance, when we view an object on the horizon). But why should removing just one of the many available depth cues have such a transformative effect? After all, we still have access to the numerous pictorial cues that are thought to specify the depth of the scene. So treating binocular disparity as just another depth cue doesn't appear to do justice to the transformational effect of either gaining or losing disparity. Some authors have argued that although binocular disparity transforms our visual experience it doesn't affect the perceived geometry of the scene, but this position does not appear to be sustainable.

Keywords Monocular stereopsis · Monocular depth · Binocular scaling Pictorial depth · Cognitive phenomenology

So far we have considered the *interaction* between pictorial cues and binocular disparity. But what about the impression of stereopsis from pictorial cues *in the absence* of binocular disparity? If binocular disparity is just another depth cue, then why does the visual system leave us with only (a) a *significantly attenuated* impression of stereopsis when we close one eye,

---

The original version of this chapter was revised: Post-publication corrections have been incorporated. The erratum to this chapter is available at https://doi.org/10.1007/978-3-319-66293-0_5

© The Author(s) 2017
P. Linton, *The Perception and Cognition of Visual Space*,
DOI 10.1007/978-3-319-66293-0_3

and (b) arguably an *absence* of stereopsis when we view objects at far distance? And if the residual stereopsis that remains in the monocular case can be explained in *purely optical* terms (see Chap. 4), why should we think that pictorial cues are a distinct source of depth information? This debate goes to the very heart of how stereopsis ought to be conceived, and there are two positions in the literature:

According to the first position, stereopsis conveys the *perceived geometry* of the scene. In which case, we have to admit that the perceived geometry of a scene (a) becomes *significantly flatter* when we close one eye at near distances, and (b) becomes *flat*, or at least *very significantly flatter*, at far distances. For those such as myself and Koenderink et al. (2011), for whom stereopsis operates *purely* at the level of *phenomenal geometry*, and makes no claims about the *physical geometry* of the *physical world*, this is readily admitted (Koenderink claims that vision has only *subjective* meaning, see Koenderink 2011, whilst I follow Austin 1962; Travis 2004, 2013 in suggesting that vision is *silent*—it makes no claims or representations; physical, subjective, or otherwise). By contrast for Representational theories of perception, the flatness that is apparent with the loss of binocular disparity is a massive embarrassment since it represents the *physical geometry* of the *physical world* as being *significantly flatter* when we close one eye, and either *flat* or *very substantially flatter* when viewed from far distance.

It is for this reason that Representationalists such as Tye (1993) and Hibbard (2008) avail themselves of a position more closely associated with Sensationalism (Peacocke 1983) and Intentionality (Vishwanath 2010, 2014), namely that variations in stereopsis do not (or at least need not) affect the perceived three-dimensional geometry or shape of the object in question. Instead, stereopsis is conceived as an *additional quality* (or *fourth* dimension) of visual experience which can vary whilst the three-dimensional quantities of the object remain fixed. This enables Representationalists such as Tye (1993) and Hibbard (2008) to maintain that vision represents the spatial properties of the physical world as being invariant, whilst nonetheless permitting our stereoscopic impressions to vary significantly depending upon the viewing conditions. But, as I argue in this chapter, the suggestion that scene geometry can be quarantined from our stereoscopic impressions is hard to maintain.

1   Binocular Vision at Far Distances

When we view objects (such as buildings and mountain ranges) at far distances they can often appear artificially flat, almost like cardboard cutouts. According to Sacks (2006), the dome of St. Paul's Cathedral looks like 'a flat semicircle' when viewed from afar, until an optical aid is used to reintroduce binocular disparity:

> ...I made myself a hyperstereoscope.... With this, I could turn myself, in effect, into a creature with eyes a yard apart. I could look through the hyperstereoscope at a very distant object, like the dome of St Paul's Cathedral, which normally appeared as a flat semicircle on the horizon, and see it in its full rotundity, projecting towards me.

There are two observations about this passage: The first is that we experience such a demonstrably *false* perception of the dome of St. Paul's— the object itself is an almost perfect hemisphere, and yet none of its volume or rotundity is apparent in our perception. The second is that this rotundity can be reintroduced but only, it would seem, with the reintroduction of binocular disparity.

This raises the following question: Why are we reliant upon the presence or absence of binocular disparity to experience the rotundity of the dome? And why don't the other depth cues 'fill-in' the depth in binocular disparity's absence? After all, a central contention of Cue Integration is that when depth cues are seriously in conflict, the visual system acts in a *robust* manner and *vetoes* (or at least *significantly down-weighs*) any aberrant cue. And yet in this context the depth we perceive is the depth specified by the aberrant cue, namely binocular disparity.

Contrast this conclusion with Johnston et al.'s (1993) early aspirations for Cue Integration, according to which pictorial cues ought to veto aberrant binocular disparity. For instance, Johnston et al. argue that such *vetoing* occurs when we switch the images in a stereoscope, thereby placing pictorial cues and binocular disparity in conflict: we often do not experience the reversed images as reversed in depth even though this is what is specified by disparity, and Johnston et al. take such failures of pseudoscopy as evidence that a 'combination of a number of different pictorial cues, such as perspective, shading, and texture, can veto stereopsis'. By contrast, I would argue that the more complex the scene, the less chance there is a coherent inverted percept to adopt (see the potential role of overhangs in vetoing Binocular Depth Inversion Illusions (BDII), discussed in Chap. 2), which explains the failure of pseudoscopy in complex scenes.

A similar point to Johnston et al.'s is made by Rogers and Gyani (2010) in the context of Reverspectives (Fig. 4 in Chap. 1), where the eventual compromise is not just flatness (i.e. the vetoing of pseudoscopy), but the inversion of the depth specified by binocular disparity. Rogers and Gyani claim that: 'For a stationary observer, viewing a reverspective with ... two eyes shows that perspective gradients can override ... binocular disparities...'. But again, in light of my discussions of Reverspectives and the hollow-face illusion in Chap. 2, I would question the validity of this claim.

For those who reject my analysis of pseudoscopy and Reverspectives, the fundamental problem is this: If pictorial cues can veto binocular disparity in the context of pseudoscopy and Reverspectives, then why are distant objects perceived as being artificially flat? After all, the visual system only has to compensate for the absence of binocular disparity, rather than overriding it (as it has to in pseudoscopy and Reverspectives)? Indeed, in Landy et al. (2011) three of the leading advocates of Cue Integration (Michael Landy, Marty Banks, and David Knill) advance the fall-off of binocular disparity with distance as exactly the kind of concern that Cue Integration ought to address. And yet contrary to this hypothesis, when we view objects at far distance the depth specified by stereopsis still seems to reflect the aberrant depth specified by binocular disparity: Cue Integration fails where simply adding binocular disparity via a hyperstereoscope does not.

Vishwanath (2010) raises similar concerns for Cue Integration: He observes that at far distances objects lack stereopsis and look 'almost *pictorial*', and suggests that Cue Integration has difficulty explaining the absence of stereopsis in this context. Nonetheless, Vishwanath would resist my suggestion that distant objects appear *flat*: As we have already discussed in Chap. 1, Vishwanath suggests that 2D images are a perfectly adequate way of conveying the geometry of a scene, so he argues that the *perceived geometry* (or *flatness*) of the scene is a completely different question from our subjective impression of *stereoscopic* depth.

But if stereopsis does not contribute to the *perceived geometry* of the scene, what does it contribute to? For Vishwanath, the answer is *tangibility*: an anticipation or prediction of our ability to *successfully interact* with the object. So, whilst the orthodox understanding is that:

Stereopsis reflects the *three-dimensionality* of an object.

Vishwanath argues that:

Stereopsis reflects the *tangibility* of an already three-dimensional object.

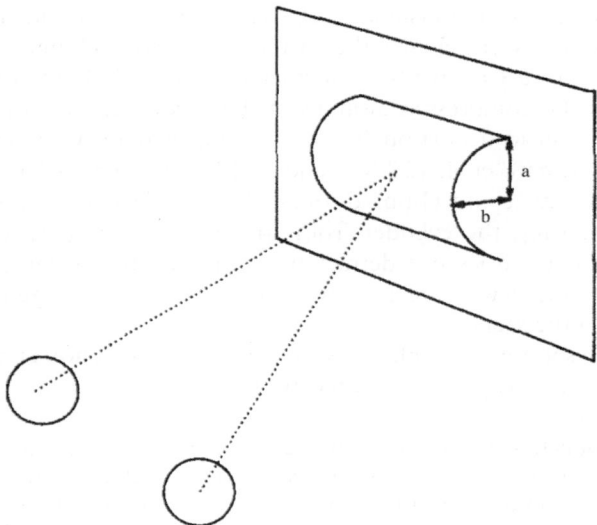

Fig. 1   Experimental set-up in Johnston (1991). From Johnston (1991). © Elsevier

But as Vishwanath observes, in order to successfully interact with an object we need reliable egocentric distance information. Consequently for Vishwanath stereopsis represents 'the perceptual presentation of the *precision* (statistical reliability) of egocentrically scaled depth information'.

According to Vishwanath (2014) this principle explains why a large voluminous object viewed from far away (such as St. Paul's Cathedral) does not induce the same stereoscopic impression as a small-scale model viewed up close: 'large absolute depth values perceived imprecisely are expected to produce a weaker impression of stereopsis than small depth values perceived very precisely...'. By contrast, I would argue that the fall-off of binocular stereopsis with distance is better understood as a distortion of shape, specifically one-half of the 'systematic distortions of shape from stereopsis' documented by Johnston (1991).

Johnston tested the perceived shape a circular cylinder at various distances (Fig. 1) and found that although subjects were close to veridical at intermediate viewing distances (107 cm) (i.e. $b = a$ in Fig. 1) at closer viewing distances (53.5 cm) the cylinder appeared elongated towards the viewer (i.e. $b > a$), whilst at further distances (214 cm) the cylinder appeared to be compressed away from the viewer (i.e. $b < a$). This pattern of geometric distortion has been confirmed by Glennerster et al. (1996), Bradshaw et al. (2000), and Scarfe and Hibbard (2013), and the fact that Scarfe and Hibbard found that this distortion could be cancelled by viewing the cylinder from an oblique angle (i.e. introducing perspective cues) does not detract from the fact that when perspective cues are absent, viewing distance is liable to distort our judgement of the geometry of the scene.

But in response to similar observations by Doorschot et al. (2001), Vishwanath (2010) presses the objection that:

> ...we routinely see pictures of objects we have never seen before and in the subsequent encounter with the real object, we seldom notice any significant difference in depth scaling (i.e. we do not have the impression that objects appear considerably flatter ... in pictures than they do in real life...)...
>
> Moreover, we can all agree on whether a depiction of a 3D object matches the real 3D object—precisely what separates good drawings from not so good ones!

But the difficulty with this approach is illustrated by amending the last quotation:

> Moreover, we can all agree on whether a depiction of a 3D object matches the real 3D object—precisely what separates good [*sculptures*] from not so good ones!

To give an example: imagine Michelangelo had asked an apprentice to fashion a small-scale replica of his *David* to be mounted on a wall and viewed only from the front. The apprentice produces a model that is a scaled replica apart from the sagittal (depth) plane where it is severely compressed. Looking at this object on the wall, Michelangelo immediately notices that the replica lacks the appropriate stereoscopic depth. Is he justified in rejecting the model? Or can the apprentice legitimately claim, following Vishwanath, that the model does in fact faithfully reproduce the geometry of the original sculpture? Now, it is tempting to suggest that the model accurately represents Michelangelo's *David* when viewed from afar, but this is no part of Vishwanath's claim for two good reasons: First, there is no such requirement for drawings or photographs—we don't have to view the object from a far before we can appreciate their accuracy. Secondly, the admission that a depiction only captures the geometry of the object at one specific distance implicitly concedes the point that I am trying to emphasise, namely the absence of shape constancy with changes in viewing distance.

But haven't I painted myself into a corner? After all, under this account a 2D photograph of Michelangelo's *David* would appear to resemble the original sculpture better than another 3D object with marginally less depth. But that's only if we see the 2D photograph as a *replica* (that appeals to our *perception*), rather than a *depiction* (that appeals to our *cognition* or *understanding*). Clearly, if the apprentice had produced a 2D painting, this would have been a worse *replica* than the compressed 3D model. But the reason the 2D image appears to function so well as a *depiction* is because it removes stereopsis from the equation: It removes the one aspect that is liable to vary with distance, and so can depict the original object at *all* distances, not just one. It is for this reason that we are not surprised when we see the real object after viewing a 2D image, even though we would be surprised if we had only ever seen the compressed 3D model.

Nor is it apparent why the distortions reported by Johnston (1991) should reflect the precision of distance information in the way that Vishwanath's account suggests:

First, from a Cue Integration perspective, it is unclear why the objects in Johnston (1991) should appear distorted at distances *closer* than 1 m (where estimates ought to be at their most precise). Given Vishwanath's rejection of *representation* in favour of *tangibility*, he might argue that the positive distortions at close distances are actually the visual system's *enhancement* of stereopsis as a kind of *accentuated* invitation to interact with the object. The problem with this response is that the subjects in Johnston (1991) would be left grasping for an elliptical object even though successful interaction with the world requires that they form a circular grasp.

Second, stereopsis is unsuitable for conveying information about precision. On the one hand, as Vishwanath (2013) explains, he is trying to get stereopsis to do the same job that is often attributed to blur: conveying the degree of determinacy with which the visual system is able to specify the location of the object. On the other hand, our subjective experience of blur and stereopsis are very different: At least in the context of blur, it could plausibly be claimed that vagueness *at the level of the visual system* is made manifest as vagueness *at the level of our visual experience*. I wouldn't make this argument, but it is something that both Tye (2003) and Vishwanath (2013) have in common. For instance, Tye argues that when we see blurrily, our visual experience 'makes no comment on where exactly the boundaries lie'. By contrast, the fall-off of stereopsis with distance appears to slowly eradicate the volume of the object in favour of increasing degrees of flatness: as distance increases, the volume of the object doesn't seem *vaguer* but *flatter*. So unlike the *roughly right* (*imprecise* but *accurate*) impression of the object's volume that blur affords us, stereopsis appears to afford us an *exactly wrong* (*precise* but *inaccurate*) impression instead.

Unlike Vishwanath (2010, 2014), Domini and Caudek (2011) recognise that our impression of objects and scenes is liable to *flatten* with distance. Indeed, they argue that the failure of shape constancy with distance is an important empirical finding that any plausible account of Cue Integration must face up to. And yet, their own solution effectively dismisses this concern: They argue that human interaction with the environment is limited by the reach of our arms. Consequently, they argue that we *do* have an accurate impression of 3D shape *when it matters* (50–70 cm), and that any distortion beyond 70 cm is simply a consequence of pushing the visual system beyond its operating window. But this response cannot explain why (a) shape distortions occur at distances *closer* than 1 m (in particular, Domini and Caudek's own optimal range of 50–70 cm), nor (b) the fact that we do not simply use vision to *act*, but also to *plan* future actions: We would not accept a visual system that made us *blind* beyond 1 m as evolutionarily sound (even though it preserves vision for the purposes of grasping); so why should we excuse the shortcomings of a visual system that only enables us to see a *distorted* representation of our environment beyond 1 m?

2   Monocular Vision at Near Distances

It is surprising the degree to which closing one eye can *flatten* a real-world scene. But we can recognise this fact without going to the other extreme of suggesting (as Sacks 2006, 2010; Barry 2009 do) that monocular vision is *entirely flat*. Instead, the fact that monocular vision often appears so much *flatter* than binocular vision, to the extent that it could be *mistaken* for being flat, is itself a fact worthy of explanation. So, too, is the *transformational*, rather than merely *supplementary*, contribution that binocular disparity makes to the volume of objects and scenes. This was a central component of Barry's (2009) transition from monocular to binocular vision. As she explains, her expectation, engendered by the Cue Integration literature, was that binocular disparity should merely *augmented* her perception of depth; after all, she had access to every other depth cue apart from binocular disparity and vergence:

> Gaining stereovision, I thought, would augment my perception of depth but not change it in any fundamental way. So, I was completely unprepared for my new appreciation of space…

So the first problem for Cue Integration is that binocular disparity can be *transformational*. Under a Cue Integration account depth perception *with binocular disparity* and depth perception *without binocular disparity* are supposed to be the very same process: Yes, the visual system has lost one of its sources of depth information, but it has plenty of others at its disposal. So, why didn't the addition of binocular disparity merely augment, rather than transform, Barry's visual experience?

The second problem for Cue Integration is *how* binocular disparity is transformational: Yes, binocular disparity may ensure that depth perception is *more accurate* and *more precise*, but there is no reason it ought to fundamentally change the geometry of the scene. Indeed, in light of Johnston (1991), I would challenge the assertion that binocular disparity is necessarily *more accurate*, leaving us merely with *improved precision*. But why should monocular vision's being less precise explain its comparative flatness? That would be like the visual system being unable to determine where in the range of 15–20 cm the depth value lies, so setting it at 5 cm instead.

In response to this discussion, one commentator suggested that Cue Integration can easily account for Barry's transformational experience: 'Isn't expanding the depth by a factor of 10 fundamental? It could make cubes into rods, tiles into cubes, etc'. But this is exactly my point: Why should there be a ×10 (or even a ×2) increase in depth in the first place? Either the accentuated depth of binocular disparity is veridical, in which case why are pictorial cues ineffective at replicating this depth in the absence of binocular disparity? Or the accentuated depth of binocular disparity is non-veridical, in which case why does Cue Integration enable it to dominate the binocular percept?

But perhaps there is a way out of this dilemma? What if, instead of thinking about stereopsis in terms of the 3D shape of an object, with an *increase in stereopsis* equating to an *increase in volume*, we think of stereopsis as a *quality* of vision: a *fourth* dimension that can vary even though 3D shape remains fixed? We can come to this conclusion in one of three ways:

1. Sensationalism: Peacocke (1983) describes closing one eye as an occasion where (a) the 3D geometry of the scene remains fixed, but where there is nonetheless a reduction in something, specifically (b) our *subjective sensation* of depth. Peacocke, therefore, describes monocular and binocular vision as two experiences that have the same *representational* content but differ in some other intrinsic respect, namely depth as a *sensational* property.

2. Representationalism: This conclusion is resisted by Tye (1993) who argues that the distinction between monocular and binocular vision can be articulated in representational terms so long as we broaden our notion of perceptual representations to include second-order representations (i.e. *representations about* our perceptual representation of the world). In this context Tye argues that the subjective difference between monocular and binocular vision is a representation of *the determinacy with which* our perceptual representation of the world is made. Indeed Hibbard (2008) hears echoes of Cue Integration in Tye's approach since Cue Integration requires not just (a) an estimate of scene properties but also (b) the visual system's confidence in those estimates.

But under Cue Integration these confidence estimates are never directly conveyed to the subject. By contrast, Hibbard suggests that stereopsis might be the means by which these confidence estimates are *directly* conveyed, with the 'vividness' of binocular stereopsis (over monocular stereopsis) conveying the enhanced reliability of binocular depth judgements. It is this commitment to stereopsis as a *quality* of visual experience, over and above the geometry of the scene, that differentiates Hibbard's account from the orthodox account of Cue Integration: 'What is at stake is the question of whether there is a relevant quality of perceptual experience beyond the geometrical representation of three-dimensional shape'.

3. Intentionality: As we have already seen, Vishwanath (2010) agrees with Peacocke (1983) and Hibbard (2008) that the 3D geometry of the scene is conveyed irrespective of variations in stereopsis. But for Vishwanath, the difference between monocular and binocular stereopsis is not epiphenomenal (as it is for Peacocke), nor simply a reflection of the reliability of binocular depth estimates per se (as it is for Hibbard), but a reflection of the precision with which the scene is scaled. So whilst Vishwanath (2010) suggests: 'I have proposed that the core quantitative aspects of cues—statistical reliability—appears in itself to be *qualitatively* expressed in perception', and this might sound superficially similar to Hibbard's account, we have to understand that for Vishwanath, Cue Integration is not concerned with 3D shape (see Vishwanath and Domini 2013), but only the scaling of 3D shape.

This leads us to the second important distinction between Vishwanath and Hibbard: Under Vishwanath's account, stereopsis represents the transition from *relative depth* to *scaled tangible depth* (with the *precision*, rather than the *magnitude*, of the scaled depth determining the degree of stereopsis: 'large absolute depth values perceived imprecisely are expected to produce a weaker impression of stereopsis than small depth values perceived very precisely...'). By contrast under Hibbard's account stereopsis is no longer a *spatial* quality since it no longer represents *depth*; geometric, tangible, or otherwise. As Vishwanath (2013) explains:

> I argue that stereopsis is the phenomenal sense of precision with which we know the location of objects in the light of wilfully acting upon them.... This is in direct opposition to [Hibbard (2008)'s view] that it is merely an indicator of how accurately and precisely the so-called depth cues—in combination—specify the Euclidean structure of the scene.

But Vishwanath's articulation of stereopsis exclusively in terms of depth scaling brings its own concerns: Binocular disparity is pre-eminently a cue to relative depth, so it seems surprising that under Vishwanath's account binocular disparity only contributes to the precision of our estimates of scale and not to the precision of our estimates of the 3D layout of the scene. Indeed, under Vishwanath's account binocular disparity simply becomes a proxy for vertical disparity (Rogers and Bradshaw 1993, 1995).

1. 3D Shape: But what of the claim that is common to all three accounts, namely that the 3D layout of the scene is unaffected by the

subjective difference between monocular and binocular vision? In Sect. 1, I argued that the absence of *shape constancy* put pay to the suggestion that viewing distance and 3D shape are distinct, but is this account any more sustainable in the context of the distinction between monocular and binocular vision? Do we at least experience shape constancy when we close one eye? The classic study that suggests otherwise is Thouless (1931a, b). Thouless asked his subjects to estimate the ratio of the y-axis (height) to the x-axis (width) of circles slanted away from the observer. He found that once familiarity cues had been removed, subjects' monocular judgements were close to the perspective projection. By contrast, in binocular vision he found a 'phenomenal regression towards the real object': Whilst subjects could not differentiate a monocularly viewed circle slanted at 30° from a ½ height ellipse viewed front on, they could when the slanted circle was viewed binocularly.

Indeed, what is startling about these results is that they *underreport* the phenomenon: The question that interests us is the question of *shape constancy*, i.e. whether a circle can *look circular* with binocular viewing, but *look elliptical* with monocular viewing. This is an all-things-considered judgement that takes account of the x-, y-, and (most importantly) z-axis. By contrast, Thouless was preoccupied by the fact that a slanted circle casts an elliptical retinal image, and he wanted to know whether what we saw was simply this retinal image? Hence, he only asked his subjects to perform an x-axis versus y-axis comparison. Now, it is a startling fact that the shape *in*constancy of monocular versus binocular vision shows up in the relationship between the x- and y-axis (as Elner and Wright 2015, observed when they replicated Thouless' experiment, they had one participant who couldn't believe that they hadn't performed a perspective match). But this *isn't* the question that interests us. For instance, we wouldn't think that a perspective match (solely in terms of the x- and y-axis) would be a good way to evaluate binocular shape constancy: It would have eradicated the results in Johnston (1991) where the task was to compare the z-axis with the y-axis. Similarly, I would argue that the effect of closing one eye is primarily experienced as a reduction of depth in the z-axis, and that what really interests us isn't the fact that a slanted circle *looks more circular* when viewed binocularly (Thouless' phenomenal regression), but the fact that it can actually *look circular*, i.e. it can appear to have an equal radius. Often when looking at a slanted circle *binocularly*, it is relatively easy to convince yourself that there is a constant radius thanks to the depth in the z-axis. By contrast, it is surprisingly difficult to convince yourself that there really is a constant radius when looking at a slanted circle *monocularly*, and it is in this sense that monocular vision *flattens* perception.

Still, the claim that closing one eye flattens the shape of an object is resisted in some quarters, even in the context of Thouless' slanted circles. For instance, Schwitzgebel (2006) asks if we are 'willing to say that a circle viewed at an angle looks like an ellipse monocularly but not binocularly? To contemporary sensibilities this may seem strange'. But this only seems strange if we conflate the *phenomenal geometry* of vision with the *physical geometry* of the physical world. Of course it sounds strange to suggest that closing one eye makes 'the world' seem flatter if we mean the *physical geometry* of the physical world. But vision scientists are increasingly willing to distinguish the *phenomenal geometry* of vision from the *physical geometry* of the physical world: First, and foremost, because of the apparent absence of binocular shape constancy with distance, see Johnston (1991). Second, in order to question whether *phenomenal geometry* obeys the very same laws of Euclidean geometry that *physical geometry* is supposed to obey, see Koenderink et al. (2010). Third, to incorporate the disassociation between location and shape that is possible in *visual space* but not *physical space*, see Loomis et al. (2002). And finally, to account for the apparent disassociation between pictorial and physical depth, see Koenderink et al. (2011).

But if closing one eye is liable to distort the perceived geometry of a scene, then why isn't this distortion always apparent? For instance, Volcic et al. (2014) cite a number of studies that suggest that our perception of shape is not affected by variations in stereopsis: Vishwanath and Hibbard (2013), Wijntjes and Pont (2012), Cooper and Banks (2012), and Erkelens (2013). But as Erkelens (2013) himself recognises, the truth is more complex:

> Perceptual effects of binocular disparity are prominently present in certain experimental paradigms [depth from Random-Dot Stereograms (Julesz 1960), optimal cue-combination (Knill and Saunders 2003), binocular slant rivalry (van Ee et al. 2002)], less influential in others [depth from reverspectives (Papathomas 2002), hollow-mask illusion (Hill and Bruce 1993)], and absent in still others [perceptual ambiguity in Necker cubes (Erkelens 2012), Schroeder staircase (Beer 1990), slant in perspective-rich stimuli (this study)].

> The results of the present study suggest that depth from disparity is not included if the monocular percept is dominated by linear perspective. Why binocular disparity is included in some cases and not in others is still a mystery.

But this is only a mystery if we fail to take the lesson of Todd and Norman (2003) seriously. There, you'll remember, subjects *judged* that there was more depth in a monocular motion display than in a binocular disparity plus motion display; even though when they closed one eye whilst viewing the binocular disparity plus motion display (turning it back into the monocular motion display), they saw *less* depth. Todd and Norman took this as evidence that our *perception* and *cognition* of depth can come apart, and there are good reasons to believe that this is exactly what is going on in the cases that concern Erkelens (2013).

First, take the Reverspectives and hollow-face illusion that Erkelens mentions. As we discussed in Chap. 2, there is every possibility that what is going on here is a disassociation between our *perception* and our *cognition* of depth.

Second, Erkelens' proposal, namely that binocular disparity only matters in the absence of linear perspective, can't explain the difference between the monocular and binocular viewing of Reverspectives. As Erkelens himself admits, the illusory impression of inverted depth persists at closer distances for monocular viewing even though the Reverspective is dominated by perspective cues.

Third, Erkelens' halfway house between my position on the one hand, and Peacocke, Hibbard, and Vishwanath on the other, seems to be conceptually unstable: The problem is what does binocular disparity contribute to under Erkelens' account when it *doesn't* contribute to the perceived geometry of the object? Something clearly changes *visually*, but what is that something?

I would argue that those studies (such as Erkelens 2013) where our determination of shape is unaffected by binocular disparity are best understood as reflecting the fact that we are able to disassociate our *cognition* of shape from our *perception* of shape. Indeed, when you think about how binocular disparity falls off with distance, there are good evolutionary reasons for being able to decouple our *cognition* from our *perception*: We want to be able to *understand* the geometry of the scene as invariant with distance, even if our *perception* is telling us quite the opposite.

The possibility that our *cognition* of shape might outstrip our *perception* of shape provides Howard and Rogers (2012) with good reason for supposing that the mechanisms that compensate for the failure of shape constancy with distance (Johnston 1991) need not be perceptual:

> None of the failures in depth constancy revealed in the laboratory seem to be evident in the real world. In a laboratory, subjects are well aware that objects are not constant. This is obvious to them because they are asked to adjust the 3-D structure of test objects. In the real world we can usually safely assume that objects remain the same at different distances. Also, we can usually see several similar objects, such as people, houses, or cars, at different distances at the same time. With such reliable assumptions and rich displays we do not need refined mechanisms for distance scaling of disparity.

If *cognitive* mechanisms can explain the impression of binocular shape constancy with distance, they might equally explain the impression of shape constancy between monocular and binocular vision.

2. Reliability: So much for the *negative* dimension of the above accounts, namely that stereopsis does not affect the perceived 3D shape of objects. What about the *positive* dimension of Hibbard's (2008) and Vishwanath's (2010) accounts, namely that stereopsis must in some sense track the *reliability* of our depth percepts?

a. Object Shape: Let us begin with Hibbard's (2008) claim that the *plastic effect* (the increased determinacy of 3D shape with binocular vision) reflects the improved precision of our 3D shape estimates. In agreement with this hypothesis Scarfe and Hibbard (2013) found that when strong perspective cues are present the only contribution that binocular disparity appears to make is to the *precision* rather than the *accuracy* of our determination of shape: Monocular observers were as *unbiased* as binocular observers in determining shape, but they were only able to estimate shape with half of the *precision* of binocular observers. Nonetheless, these findings only apply when strong perspective cues are present, and Scarfe and Hibbard (2013) admit that binocular stereopsis is liable to distort the perceived shape of objects when perspective cues are absent. Consequently, Scarfe and Hibbard are heavily reliant upon the presence of perspective cues to quarantine this effect, which ought not to arise under Hibbard's (2008) account.

b. Scene Layout: What of the other claim of Hibbard's (2008) account, namely that the *coulisse effect* (the increased vividness of the separation between objects with binocular vision) necessarily reflects the increased precision of our estimates of depth separation with binocular vision? Well, Hornsey et al. (2015) recently assessed the extent to which binocular disparity contributes to our judgements of ordinal depth in stereo-images. They chose this task as it promised to highlight the clear benefits of binocular viewing. And when subjects were asked to make ordinal depth judgements *along the surface of an object* it did, confirming Scarfe and Hibbard's (2013) findings that binocular disparity improves the precision of judgements of 3D shape.

By contrast, Hornsey et al. found no improvement in ordinal depth judgements of points *on separate objects*. Indeed, rather paradoxically, the threshold for determining the ordinal depth between two points on separate objects was *higher* for binocular vision than it was for monocular vision. But if this is the case, and binocular vision does not improve the reliability of our ordinal judgements for points on separate objects, then Hibbard's (2008) account cannot explain the *coulisse effect* (the stereoscopic impression of the separation between objects) even though (a) a common complaint with Victorian stereograms is that the *only* contribution they make is to this *coulisse effect*, see Koenderink et al. (2013, 2015b), and (b) the *coulisse effect* was one of the defining features of Barry's (2009) transition from monocular to binocular vision.

c. Egocentric Distance: What about Vishwanath's (2010) claim that binocular stereopsis reflects the improved precision of our egocentric distance estimates?

On the one hand, binocular disparity does seem to improve the precision of our *near* distance estimates. To give an illustration, in the 1990s one of the leading explanations for the grasping deficits of amblyopic (lazy eye) subjects was that since they were effectively monocular they must be *underestimating* near distances, see Servos et al. (1992), Servos and Goodale (1994), and Goodale and Servos (1996). But when this hypothesis was tested in the early-2000s, Servos (2000) and Loftus et al. (2004) found that monocular subjects were no less *accurate* at estimating near distances; instead, the only benefit that binocular disparity appeared to provide was *precision*: monocular judgements were more variable, and therefore less reliable.

But when it comes to *intermediate* distances the evidence points in the other direction: Ooi and He (2015) asked subjects to estimate the distance of an object placed on the floor of a well-lit corridor between 2.7 and 6.9 m. Not only did they find that (a) monocular estimates were just as *accurate* as binocular estimates, they asked (b) 'Can the visual system use binocular depth information to improve the precision of its performance?', but found no significant difference between the *precision* of monocular and binocular estimates. Indeed, even when the objects were suspended in free space, thereby removing the ground-plane as a depth cue, this change primarily affected the *accuracy* of monocular distance estimates (with monocular subjects underestimating distances > 5 m) rather than their *precision*.

3   PICTORIAL DEPTH

But if I am right, and stereopsis is inseparable from the 3D geometry of the scene, then how are we to make sense of *pictorial depth*? After all, Vishwanath (2010) suggests that pictorial depth is the perception of 3D geometry *without stereopsis*. But for Koenderink (1998) the fact that stereopsis is inseparable from 3D geometry does not mean that we have to give up on pictorial space. Instead, we simply recognise that pictorial depth is a *weak form* of stereopsis; i.e. the difference between a '2D' pictorial image and a '3D' stereo-image is merely *quantitative* rather than *qualitative*. Indeed, Doorschot et al. (2001) argue that the contribution that disparity makes to depth from shading in a '2D' image merely reflects the fact that we have two depth cues rather than one, and can be modelled by linear Cue Integration. By contrast, Vishwanath rejects both (a) a linear Cue Integration account of pictorial depth (see Vishwanath and Domini 2013) and (b) a linear Cue Integration account of the transition from pictorial depth to stereopsis (see Vishwanath and Hibbard 2013).

But the disagreement between Koenderink and Vishwanath is more fundamental and can be summarised by two separate questions:

1. Is visual depth possible without stereopsis?

2. Is pictorial depth stereopsis?

And there are three distinct positions that one could take (see Table 1).

Now I agree with Koenderink on one of these questions, namely that there is only one kind of visual depth, specifically *stereopsis*. But I agree with Vishwanath on the other question, namely that pictorial depth

Table 1   Three interpretations of pictorial depth

| | | Is pictorial depth stereopsis? | |
		Yes	No
Is visual depth possible without stereopsis?	Yes		Vishwanath
	No	Koenderink	Linton

isn't stereopsis. But simply by agreeing with Koenderink on one question, and Vishwanath on the other, I find myself endorsing a *purely cognitive* account of pictorial depth; a position that both Koenderink et al. (2006) and Vishwanath (2010) malign as trying to 'explain away' pictorial depth. As Vishwanath (2010) observes, pictorial space is thought to:

> ...belie such simplistic, 'cognitivist' explanations, and current scientific research in visual perception rarely entertains them; seeking, rather, to understand picture perception phenomena from the viewpoint of the working of visual mechanisms. (see Cutting 2003; Hagen 1980; Koenderink and van Doorn 2003)

Koenderink, Vishwanath, and I are all in agreement that, as Koenderink et al. (2013) observe, 'one may hardly assume that the viewers of holiday postcards (say) have no impressions of pictorial space, but are only aware of planar patterns of colored patches'. The only disagreement is what exactly this *awareness* constitutes:

a. Stereopsis (Koenderink): What is interesting about Koenderink et al. (2011) is that they don't just articulate pictorial depth as an instance of stereopsis, but instead treat pictorial depth as the central case of stereopsis through which we come to understand stereopsis in general as 'optically constrained awareness'. Furthermore, they treat static monocular perception of the real world as simply an application of their pictorial depth account. Indeed, for Koenderink et al. (2015b), the key distinction is not between *pictorial* depth and *real-world* depth, but whether the observer's perspective is *fixed* or not.

But it would be difficult for Koenderink et al. to maintain that *all* pictures induce, or even require, an impression of stereopsis. This is no part of Koenderink et al.'s claim. Instead, Koenderink et al. (2015a) distinguish between those pictures that induce an impression of stereopsis, and those that do not. For instance, Koenderink et al. suggest that there is no impression of pictorial space in Fig. 2a: 'Notice that 2D pictures may permit various 3D inferences even when there is no such thing as "pictorial space"'. But I would resist this conclusion, because how could we explain in purely two-dimensional terms: (i) the implied ground-plane (objects higher up in the image plane are seen as further away), (ii) the familiar-size effects (the people are too big, implying distance between them and the house), and (iii) the size-constancy effects (see Fig. 2d where the adult male looks like a giant when superimposed in the doorway of the house). I would therefore argue that the essentials of pictorial space are already there in Fig. 2a, and so whatever explains the transition from 2D to 3D space in Fig. 2a ought equally to explain the transition from 2D to 3D space in the case of, e.g. a holiday postcard.

Indeed, I wouldn't even accept Koenderink et al.'s suggestion that the occlusion in Fig. 2b is merely a 2D inference: 'No need for 3D phantasms, and a simple 2D algorithm achieves this'. Instead, I would argue that Fig. 2b implies *3D order* without *3D depth*. To give one of the most commonplace but often overlooked instances of occlusion, consider the words on this page: they are *on* the page, they *occlude* the page behind them, and yet there is no sense this implies *depth* rather than *order*. Still, this is not a mere 2D inference: Trying to see the words on a computer screen as simply black pixels *alongside* white pixels, rather than *on top of* white pixels, is surprisingly difficult.

It is true that Fig. 2c has shading, whilst Fig. 2a and b do not. Is shading the missing element that translates 2D inferences into 3D depth? Koenderink et al. are open to the possibility: 'The case of Figure [2c] is perhaps different. It has been used often in vision research (Metzger 1975; Ramachandran 1988). Textbooks tell you that you are supposed to "see a sphere," that is a pictorial, volumetric object'. But such a position would be hard to square with Koenderink et al.'s (2015c) account of depth from shading, according to which shading has nothing to do with inverse optics but is instead merely the *articulation* of relief (in much the same way that contour lines on a map articulate relief). Indeed, Koenderink et al. regularly use contour lines as an intermediate step between local judgements and global surface (see Fig. 3b), and yet no one would suggest that we see Fig. 3b in stereoscopic depth: It is a 2D line arrangement that we *interpret* as having 3D meaning. So, there doesn't appear to be anything inherently stereoscopic about either (i) pictorial space or (ii) shading under Koenderink et al.'s account. Indeed, the point can be illustrated even more clearly with an image that

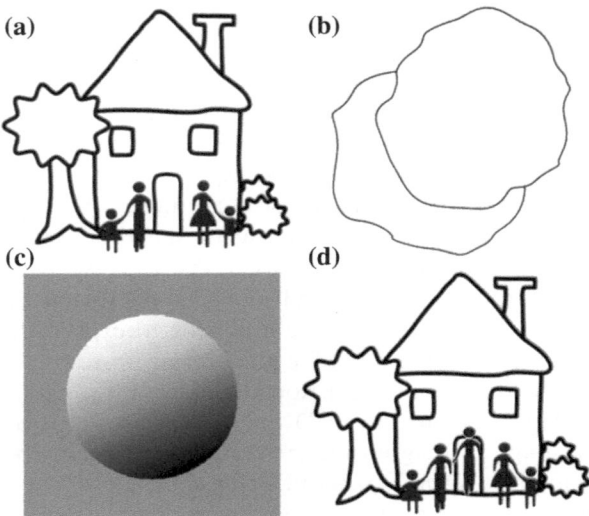

Fig. 2 Three examples of pictures **a–c** from Koenderink et al. (2015a), with **d** which is a variation of **a** but with the adult male figure placed in the doorway to demonstrate size constancy. From Koenderink et al. (2015a). CC-BY-3.0

is in between Fig. 2a and b, on the one hand, and Fig. 2c, on the other hand, such as Gregory's dalmatian (Fig. 4): Unlike Fig. 2c, there are virtually no luminance gradients indicating shading. Instead, there are just black and white patches. On the one hand, once you recognise the scene for what it is, it has a rich pictorial space: To borrow from Koenderink et al. (2013), one may hardly assume that viewers are only aware of black and white patches. But on the other hand, stereopsis appears to be completely absent in this case.

Furthermore, if the impression of pictorial relief protruding or receding from the page is truly stereoscopic, then we should be able to say *by how much* the relief protrudes or recedes from the page: either in centimetres or, at the very least, by specifying the separation with our thumb and forefinger. Often observers who are adamant that they see depth in pictures equivocate when the question is posed in these terms. But Koenderink et al. (2015a) are quick to dismiss this aspect of stereopsis, insisting that: 'pictorial space and the space the observer moves in (Koenderink et al. 2004; von Hildebrand 1893) are categorically

(a)   **(b)**   **(c)**

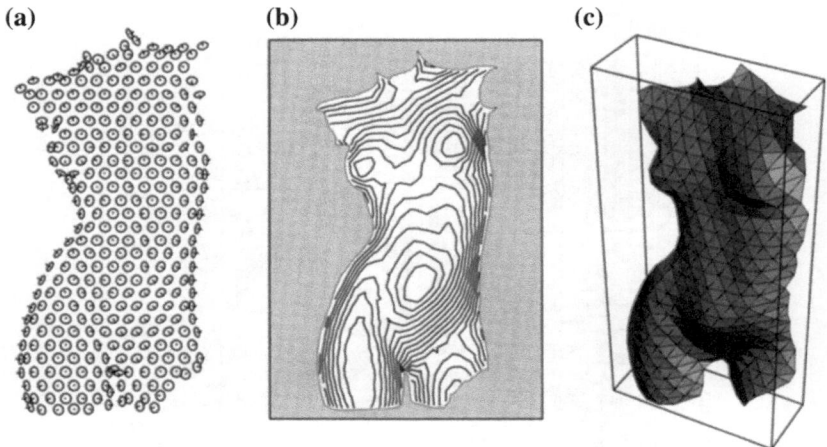

Fig. 3 Koenderink's (1998) illustration of a depth map built up out of **a** local judgements by subjects using a Tissot gauge, **b** a contour map obtained by integrating these local judgements, and **c** a 3D render. From Koenderink (1998). © The Royal Society

disjunct'. But is this correct? When we watch a 3D movie it is not uncommon to see objects *protruding* from the screen towards us or even *floating within our grasp*, whilst horizons and chasms look like they *recede* many hundreds of feet behind the screen. And yet it appears that Koenderink et al. in no sense want to preserve this aspect of stereoscopic viewing, where pictorial objects appear to *inhabit* the world; choosing instead to focus on purely *geometric* relationships *within* pictorial space.

Koenderink et al. (2015a) may object that under their account pictorial space is different from a 3D movie: They argue that the pictorial surface represents a fronto-parallel plane *behind which* the depth of pictorial objects is located, so there is no *protrusion* of pictorial objects into the real world. However, first, this is not how the addition of binocular disparity to an image is experienced. Yet, binocular disparity is meant to only *accentuate* rather than *transform* the depth of pictorial objects, see Doorschot et al. (2001). Second, this is not how the depth of pictorial objects is experienced: To the extent that a Necker cube appears to have volume, its volume appears to be located *around* the pictorial plane not *behind* it. Third, the essential point still stands even if we focus

Fig. 4 Image of a dalmatian from Gregory (2005). © The Royal Society

on *recessions* in depth rather than *protrusions*. Stereoscopic depth can give the impression of objects receding many hundreds of feet from the viewer and yet there is no place for such descriptions under Koenderink et al.'s (2015a) account.

b. Unscaled Depth Relations (Vishwanath): By distinguishing between pictorial space and stereopsis Vishwanath (2010, 2014) also keeps *pictorial* space and *real-world* space categorically disjunct. But if, according to Vishwanath, pictorial space is *not* stereopsis, but is still *visual space*, then what else can it be? Vishwanath's (2014) solution is to suggest that we *perceive un*scaled depth relations. But this is equivalent to saying that we don't perceive *actual* or *instantiated* depth (after all, *instantiated* depth is stereopsis), but merely depth relations *in the abstract*; what you might call *potential* or *proto*-depth. Make no mistake, for Vishwanath, this is still depth perception properly so-called. But for me, trying to perceive *depth relations* in the abstract without any *specified* depth is like trying to perceive a colour in the abstract without any specified shade. So this is my first concern with Vishwanath's account: I know what it is to *understand* depth relations in the abstract, but I don't know what is it to *perceive* depth relations.

My second concern is essentially a restatement of the claim in Sects. 1 and 2 that stereopsis *does* go to the question of relative depth, and so the strict demarcation between 3D geometry and stereopsis that is necessary for Vishwanath's account of pictorial depth appears to break down. It is true that some of the evidence for the influence of stereopsis on geometry is based on *local* depth judgements (for instance, the Tissot gauges of Doorschot et al. 2001, see Fig. 3a), to which Vishwanath has a response (see my next concern), but it is equally true that much of the evidence is based on *global* judgements of shape (see, for instance, Johnston 1991).

My third concern relates to the way in which Vishwanath (2014) attempts to reconcile his account with Doorschot et al. (2001). Vishwanath's contention is that he and Doorschot et al. are simply talking past each other since they are concerned with different evaluations: He is concerned with evaluating *global shape* whilst Doorschot et al. with evaluating *local shape*. Vishwanath appeals to the fact that:

> Different depth judgment tasks are thought to access different aspects of depth representation or levels of processing (Glennerster et al. 1996; Koenderink and van Doorn 2003). For example Di Luca et al. (2010) found a lack of internal consistency among judgments of curvature, local slant, and depth separation between points on the same surface.

This makes a lot of sense as a suggestion about *cognition*, namely that we are liable to come to inconsistent *judgements* about what we see. But again I struggle to make sense of it as a claim about *perception*, namely that we literally *see* incompatible local and global shape. Vishwanath agrees with this point. The only reason he adopts this reconciliation between his position and Doorschot et al.'s is his insistence that Doorschot et al.'s results must operate merely at the level of *judgement*; i.e. that adding binocular disparity affects only our *judgement* of its local shape, but not how we *see* it.

My fourth concern is that Vishwanath's (2014) response to Doorschot et al. relies upon there being a *global shape percept* in the first place. But if there were such as *global shape percept* in pictures then subjects should simply be able to 'read off' the ordinal depth of any two points on the surface of a pictorial object. And yet, as Koenderink and van Doorn (1995) and Koenderink et al. (2015b) demonstrate, subjects perform surprisingly poorly at this task. This is especially surprising because, as Fig. 3b and c demonstrate, such global depth maps can be built up out of subjects' own local depth judgements. As Koenderink et al. (2015b) observe: 'This is remarkable because one could do better than the observer by using the data obtained in one experiment involving the same observer!'

Koenderink et al. (2015b) suggest that the solution to this dilemma is to recognise that subjects do not appear to have access to a *global* depth map, but merely *regions* of depth delineated by the bulges and/or ridges of the surface. What this entails for Koenderink et al. (2006) is a complete revision in the way we think about pictorial space:

> ...'pictorial space' is a catchy term to denote something that is unlikely to be a 'space' in the usual geometrical sense at all. It is hard to say what a useful formal description might eventually look like. We would guess something like a number of formal spaces with various degrees of inner coherency and only weak mutual interactions.

Similarly Koenderink et al. (2015b) are led to the conclusion that the part-whole distinction is *fundamental* (in their words, 'very basic') to pictorial depth. Such empirical observations undermine Vishwanath's (2014) appeal to the fact that in Fig. 3a and b, 'spatial integration of local judgments reveal an internally consistent *global* 3-D shape': It is true that *we* can integrate subjects' local judgements, but Koenderink et al. (2015b) would argue that this tells us nothing about what *they* actually see.

More fundamentally, the findings in Koenderink et al. (2015b) also challenge the assumption (common to both Koenderink et al. and Vishwanath) that what is going on in pictorial depth is *perception* rather than *cognition*. Koenderink et al.'s results imply that there is no *global* percept, only a collection of 'weakly synchronized patches'. But a collection of 'weakly synchronized patches' is explicitly not what we *see* when we view Koenderink et al.'s stimulus (Fig. 5).

Indeed, given that Koenderink et al. typically emphasise the phenomenology of perception, it is telling that they choose to frame their results largely in terms of fragmentary 'mental representations', 'formal spaces', and 'data structures' rather than a fragmentary visual experience. So, too, is the fact that it has taken Koenderink et al. over 20 years to put their

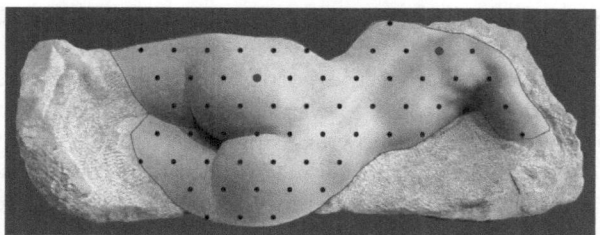

Fig. 5 Stimulus from Koenderink et al. (2015b). The task is to pick which of the two dots highlighted in *red* is closer. From Koenderink et al. (2015b). CC-BY-3.0. Based on 'Reclining Nude I' by Andrew Smith. For more information, see http://www.assculpture.co.uk/. Original image © Andrew Smith

finger on the *regional* nature of pictorial depth; something that ought to be immediately apparent if we could literally *perceive* the depth in question. By contrast, I interpret Koenderink et al.'s (2015b) results as suggesting that although we *see* the 2D image as a whole, we can only consistently *attribute* 3D depth to one region at a time: Our *attribution* of depth is piecemeal, even if our *perception* is not.

c. Cognitive Phenomenology (Linton): Suggesting that pictorial depth is purely cognitive might sound like a non-starter, but we have to remember that both Koenderink and Vishwanath are committed to *cognition* taking on at least some of the significance traditionally attributed to *perception*. In Koenderink's case, we can point to Fig. 2a. In Vishwanath's case, we can point to Vishwanath and Blaser (2010) on the ability of familiar-size cues to transform our experience: They recognise that photographs of architectural models taken with a narrow aperture (and therefore a large depth of field) enable familiar-size cues to disambiguate distance and size. Nonetheless, Vishwanath (2014) maintains (rightly in my opinion) that scale from familiar size is not perceptual, but merely a cognitive inference (see Gogel 1969; Gogel and Da Silva 1987; Predebon 1993). But if the cognitive inference of *scale* can have such a transformative effect on our experience, why not the cognitive inference of *geometry?* In short, why should we think that in order to transform our experience, this transformation has to be *perceptual?*

By contrast, if you want to maintain that our experience of pictorial space is perceptual, then you have to be prepared to point to something *visual* about it. To take one example, what changes *visually* when a Schröder Staircase switches in depth? (Fig. 6).

It might be claimed that the corner that originally appears to be behind now appears to come forward in space? But can you tell us by how many centimetres? Or show us using your forefinger and thumb. If not, what does 'come out in space' mean? I'm not being obtuse: I recognise there is some *phenomenal* difference, but I am trying to point to the fact that this phenomenal difference does *not* appear to be *visual*.

But what else could it be? Well, consider the words on a page: There is clearly a phenomenal difference between seeing something as a *mere mark* and seeing it as a *word*. Once you see something as a word its meaning immediately leaps out at you. Indeed, it no longer becomes possible to see it as an assemblage of random marks. But again *visually* speaking what has changed? Nothing. Instead, this *additional* and *purely cognitive* meaning appears to exist as something over and above our visual impression: something that does not *amend* or *replace* our visual impression, but merely *gives meaning* to what we see. And I believe something similar is going on in the pictorial case: perspective and shading are just languages that we have to learn in order to interpret marks on a page (Gibson 1966 makes a similar analogy by coupling pictures and writing together under the rubric of 'the structuring of light by artifice', but he did not take this analogy to its logical conclusion).

This analogy helps us to understand how our *experience* of a visual stimulus might change without there being a change in our *visual* experience. Like our experiences of words, I would argue that our experiences of pictures are instances of 'cognitive phenomenology': the experience of *understanding* something *in a certain way*. Although 'cognitive phenomenology' is increasingly being explored in the philosophical literature (see Bayne and Montague 2011; Smithies 2013; Chudnoff 2015) it hasn't to my knowledge been employed before to explain pictorial depth. Instead, it is often assumed that pictures are just another example of the *visual* phenomenology with which *cognitive* phenomenology is to be contrasted. Take, for instance, Bayne and Montague's (2011) insistence that 'acts of conceptual thinking are just as much experiences as percepts and pictorial representations'.

Nonetheless, my account of reading is an elaboration of the *phenomenal contrast* explored by Strawson (1994):

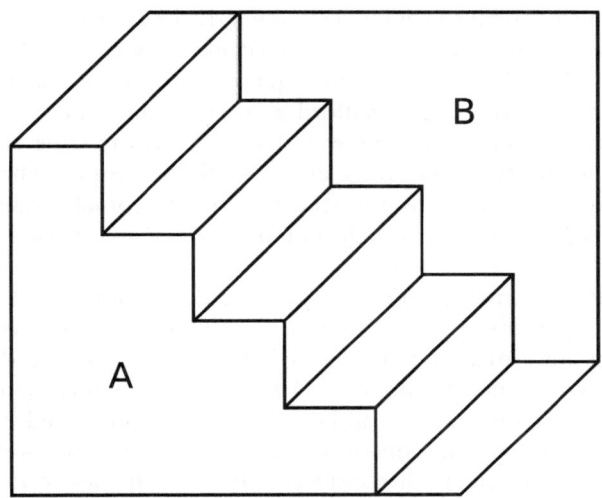

Fig. 6 Schröder Staircase. © Stevo-88. Wikimedia Commons. Public domain. https://commons.wikimedia.org/wiki/File:Schroeder%27s_stairs.svg

Philosophers will ask whether there is really such a thing as understanding-experience, over and above visual experience, auditory experience, and so on. Behind their questioning there may lie a familiar doubt as to whether there is anything going on, experientially, that either is or necessarily accompanies the understanding. The question may be asked: does the difference between Jacques (a monoglot Frenchman) and Jack (a monoglot Englishman), as they listen to the news in French, really consist in the Frenchman's having a different *experience*?

Translating this auditory example to vision, I would argue that Jacques and Jack have a very different *cognitive* experience when they read the words in a French newspaper, even though they have the same *visual* experience. Furthermore, both would have a very different *cognitive* experience from a third person who had never encountered writing, and simply saw the words as marks on a page.

Indeed, this latter case is exactly analogous to Sidney Bradford's experience of *pictures* when his sight was restored (see Gregory and Wallace 1963): He understood capital letters and clock-faces because they had already been taught to him by touch, but as Gregory (2004) recounts pictures looked flat and meaningless. Similarly for another formerly blind patient (Mike May) Necker cubes looked like 'a square with lines' (see Fine et al. 2003). And yet, I would suggest that we have exactly the same *visual* experience as Sidney Bradford and Mike May when we look at a Necker cube, in much the same way that Jack has the same *visual* experience as Jacques when he looks at a French newspaper. The only difference is that we unconsciously attach a *post*-perceptual meaning to the Necker cube, and so cannot help but **understand** the Necker cube as a 3D object.

This *cognitive* interpretation of pictorial depth not only helps us to explain how perception and cognition can come apart in Reverspectives and anamorphic images, but it also helps to capture the dual-aspect nature of pictorial depth: the sense that pictorial depth is there, but not *really* there. This dual-aspect can be made more pronounced if we introduce positive stereoscopic distortions into the image via binocular disparity: We are no longer left considering how the flatness *perceived* from stereopsis relates to the depth *conceived* from the pictorial cues, but how two very different depth structures can both be accessible to us at the same time, one via perception and one via cognition.

This point is best illustrated by Koenderink (2015), who introduces a stereoscopic bulge into a 2D image in order to test whether (consistent with his account) this bulge can 'be "captured" by the contour of a foreground object in a picture?' He concludes:

> The scene looks like nice 3-D to me.... The 3-D doesn't look much like the bulge ... at all, although it 'should' according to inverse optics. It looks more like a regular stereo presentation. Are you surprised? Why?

I disagree with Jan Koenderink on this example, so he has very kindly provided some illustrative examples with various degrees of distortion for readers to evaluate (see Fig. 7). In contrast to Koenderink, my impression of the stereo-images in Fig. 7 is of two distinct layouts: the layout specified by disparity (a Gaussian bulge) and the layout specified by pictorial cues (a woman in front of receding train tracks). I don't get an impression of the contours of the woman's body capturing the stereo-bulge: her head, etc. remain stitched to the background behind her (disparity doesn't contribute to the *coulisse effect*, i.e. a separation in depth), and her face and body are distorted by the stereo-bulge rather than filled-in in depth (disparity doesn't contribute to the *plastic effect*). The disconnect

Fig. 7 Introducing a Gaussian bulge defined by disparity into an image, with ×2, ×4, and ×8 the disparity in subsequent images. The *left* and *centre* images are for cross-fusion, and the *centre* and *right* images are for parallel fusion. Courtesy of Jan Koenderink, and based on the principles described in Koenderink (2015). © Jan Koenderink. Original image by Lisa Runnels. CC0 (Public Domain). *Image source* https://pixabay.com/en/girl-pose-squatting-young-female-788815/

between stereo-bulge and the pictorial layout is even more pronounced if we reverse the stereo-images, leaving us with a concave bulge.

Christopher Tyler has suggested that in Koenderink's (2015) original image the conflict between the pictorial cues and the *concave* bulge (from cross-fusing the centre and right images) is immediately apparent, whilst the conflict between the pictorial cues and the *convex* bulge (from cross-fusing the left and centre images) gradually emerges over time, so there must be some stereo capture in the *convex* case at least initially. But I do not think this is correct. I agree that we notice the conflict quicker in the concave case. But I would resist the suggestion that *not initially noticing* the conflict in the convex case is equivalent to *not initially seeing it*. It is true that some influential theorists believe that *not noticing* something and *not seeing* it are one and the same, for instance O'Regan & Noë (2001). But consider gradual change-blindness (Simons et al. 2000; Auvray & O'Regan 2003): You look at an image of a scene and over the course of a minute the background changes from blue to red, but you do not notice it because the change has been gradual. As O'Regan himself observes (O'Regan 2011):

> ...you can stare at the picture endlessly and not discover the change. It is particularly noteworthy that 'you' may even be looking directly at the thing that is changing and still not discover it.

But subjects have a visual experience of the background throughout the course of the experiment. If *noticing* a change is a prerequisite for *perceiving* a change, then subjects should still perceive the background as blue at the end of the experiment. But this seems hard to maintain: if the subjects reported the colour at the end of the experiment they would report red. Nor does it seem plausible that the background suddenly switches from blue to red as they report it.

In any case, the conflict between the *concave* stereo-bulge and the pictorial cues is immediately apparent, and suffices for my purposes: The fact that we are able to disambiguate pictorial depth from the concave bulge specified by stereopsis should not be possible under either Koenderink's or Vishwanath's accounts since they both argue that stereopsis and pictorial depth exploit the very same *perceptual* mechanisms, e.g. Vishwanath (2010): 'I suggest that pictorial depth is indeed a predictable outcome of the normal visual processing of surface and depth (e.g. Cutting 2003; Koenderink and van Doorn 2003; Mausfeld 2003)'.

But how can pictorial depth and stereopsis exist in conflict in the same image? The only solution is to recognise that stereopsis and pictorial depth are talking past one another, and operating at two different levels: the *perceptual* (stereopsis) and the *cognitive* (pictorial depth).

4   INDUCING MONOCULAR STEREOPSIS IN A STATIC 2D IMAGE

But the strongest resistance against such a *purely cognitive* account of pictorial cues is likely to come from the burgeoning literature on inducing monocular stereopsis from a static 2D image, see Koenderink et al. (1994, 2013), Vishwanath and Hibbard (2013), Volcic et al. (2014), Vishwanath (2016), and Wijntjes et al. (2016), as well as the early-twentieth century literature that these studies revive: von Rohr (1903), Claparède (1904), Holt (1904), Münsterberg (1904), Ames (1925a, b), Carr (1935), Schlosberg (1941), Gibson (1947), and Gabor (1960).

Unlike Brunelleschi's attempts in the early Renaissance (see Kubovy 1986), the Zograscope 'craze' of the 1740–1750s (see Blake 2003), and the optical instruments derived by von Rohr in the early-twentieth century (the Verant in 1903 and the Synopter in 1907), modern methods of inducing monocular stereopsis from 2D images do not require us to *do* anything to the image. Instead, we only need to view it through a peephole, thereby obscuring the frame of the image (Fig. 8).

For those who already conceive of pictorial depth as stereopsis, this effect merely accentuates the pre-existing depth. By contrast, for others it introduces stereoscopic depth where previously there was none. For instance, Scholsberg (1941) differentiates his account from Ames (1925a, b) and Carr (1935) by stressing the 'all or none' nature of the effect: 'We do not have a simple addition and subtraction of factors, with more or less depth resulting. The perception seems to exist in two modes. In one it is still a picture. In the other mode we find objects in depth...'.

However we conceive of it, the weight of contemporary evidence is clearly in favour of monocular stereopsis from static 2D images. Nonetheless, there are some unresolved concerns:

First, there are isolated reports that don't quite fit with this account. For instance, Johnston et al. (1993) suggested that their 'monocular stimuli with depth defined only by texture do not provide the compelling sensation yielded by the stereo stimuli'. This is also reflected in the

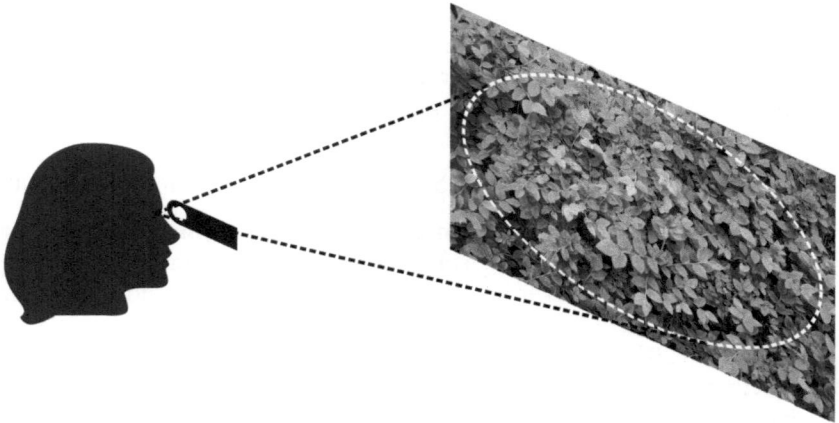

Fig. 8 Vishwanath and Hibbard's (2013) method for inducing monocular stereopsis from a static 2D image

physiological conception of stereopsis as a binocular phenomenon (see Parker 2016, discussed in Chap. 1).

Second, no one appears to have a coherent account of *when*, exactly, this monocular stereoscopic impression ought to be induced. For instance, since both Vishwanath (2014, 2016) and Koenderink et al. (2015a) agree that the effect can be induced in the presence of a visible pictorial frame, the experimental set-up in Fig. 8 is *not* required. Indeed, this would appear to cast doubt on the traditional counter-cue explanation (e.g. Schlosberg 1941), according to which monocular stereopsis from a 2D image is due to the *release* of pictorial cues from counter-cues indicating the flatness of the pictorial surface.

Vishwanath attributes the effect to the scaling of relative depth in the pictorial image using distance cues (most notably accommodation, but also the specific distance tendency and depth of field blur), but as he readily admits in Vishwanath (2016), 'the specific mechanisms underlying this effect remains to be established'. But even then, it is hard to explain why synoptic viewing (viewing two identical images side by side in a stereoscope) reportedly enhances the effect twofold (see Doorschot et al. 2001). Indeed, since stereoscopes often have lenses to set accommodation at infinity, the effect of (uncrossed) synoptic viewing under Vishwanath's account ought to be *exactly the same* as binocular vision of the real world at far distance. And yet, as we saw in Sect. 1, Vishwanath agrees with Sacks (2010) and myself that binocular vision at far distances lacks stereopsis. So something has to give. As Vishwanath (2014) himself admits, synoptic viewing is something that can only 'be partially explained' by his account.

By contrast, Koenderink et al. (2011, 2015a, b) have suggested that monocular stereopsis is simply a distinct 'viewing mode' that some observers have access to. These observers (including Koenderink, van Doorn, and Wagemans themselves) can induce the effect simply by looking at a picture and (as Koenderink et al. 2015a term it) *deploying their mental eye*. For instance, Koenderink et al. (2011) suggest that:

> This is a common experience when viewing baroque ceiling paintings. ...
> If well painted, then they easily beat any 'virtual reality' setup we have ever
> seen. The famous example (due to Pirenne 1970) is Pozzo's ceiling paint-
> ing in the church of *Sant'Ignazio* at Rome...

But Koenderink et al. (2015a) recognise that this will not chime with everyone's experience (including some of their reviewers!), so they estimate that this monocular stereoscopic viewing mode is only possible for about half the population. Ultimately, then, Koenderink et al. propose a kind of relativism: I might be right about *my* visual experience, but they are right about *theirs*. But I would regard such a relativistic conclusion as a very last resort.

Which brings me to my own preferred response, which is to suggest that the impression of depth in these cases is largely cognitive. I admit that the experimental set-up in Fig. 8 may well produce the weakest of stereoscopic effects, but this wouldn't have anything to do with pictorial cues: As I explain in Chap. 4, placing an aperture in front of the eye is liable to induce a weak concave distortion in the visual field. You can test this for yourself by pinching your forefingers and thumbs together (to make a small aperture), looking through it at a page of text on a computer screen, and moving the aperture around: The text appears subtly distorted.

Vishwanath suggests that this weak concave distortion could not account for the full phenomenology reported in the literature, and I quite agree. But I wonder whether his preferred peephole method can either? For instance, whenever optical toys (e.g. nineteenth-century paper peepshows, see Balzer 1998; Hyde 2015) sought to induce a monocular impression of separation in depth (the *coulisse effect*), they relied upon planes staggered in physical space rather than Vishwanath's peephole method (for more recent investigations of this principle, see Dolgoff 1997; Rolland et al. 2000; Akeley et al. 2004; Watt et al. 2005, and the multiplane displays by the Hua Lab: Liu et al. 2008; and the Banks Lab: Love et al. 2009, discussed in Chap. 4). So, we need to consider the basis for the claim that monocularly viewed 2D images can induce a stereoscopic impression of depth.

i. Verbal Report: Vishwanath and Hibbard (2013) and Vishwanath (2016) ask their subjects to verbally report and rank their stereoscopic experiences. I am suspicious of this method in light of the inconsistencies that Todd and Norman (2003) found between their subjects' reports (monocular depth > binocular depth) and their actual visual experience (monocular depth < binocular depth). But for Vishwanath and Hibbard (2013) this inconsistency only proves their point, since (as we saw in Sect. 2) they would argue that the stereoscopic *visual experience* in Todd and Norman is completely distinct from the *depth estimation* to which the verbal reports in Todd and Norman pertain: 'perceiving a stronger impression of stereopsis is not the same thing as perceiving a greater magnitude of depth or a different 3-D shape'. So your position in Sect. 2 will determine whether you think Todd and Norman (2003) casts doubt upon this method.

Nonetheless, even if you are tempted by this method, it is worth observing that when the subjects in Vishwanath and Hibbard (2013) were asked to pick four statements (out of a list of 9) that best encapsulated their experience, only a quarter chose 'greater separation between objects' as one of their four statements. Furthermore, whilst it is true that most subjects agreed with the statement that they experienced 'greater separation between objects' under monocular aperture viewing, it is worth noting that (a) there were subjects that strongly disagreed with this statement (in contrast to binocular stereopsis where there was no such disagreement), and (b) when asked to rate the strength of their agreement with this statement (out of '+', '++', and '+++'), the median response of the subjects as a whole was merely '+'.

ii. Reaching: Volcic et al. (2014) attempted to demonstrate just how literally the subjects' reports in Vishwanath and Hibbard (2013) could be taken: Under monocular aperture conditions subjects were liable to vary their reaching response by roughly 7 cm when pointing to different parts (near, middle, and far) of images of plants located 42 or 52 cm away.

But what does this show? Only, as Volcic et al. note, that 'under monocular viewing, responses were modulated by the different pictorial depths'. Indeed, Volcic et al. presented images of plants that were consistent with them being located in reachable space. Furthermore, Volcic et al. recognise that the fact that reaching responses were liable to be roughly the same whether the distance of the display was 42 or 52 cm suggests that there were no reliable visual or non-visual signals indicating the distance of the image plane (this should be a concern for Vishwanath's depth-scaling account). Hence, it is only natural that subjects responded to the only cue available to them, namely the familiar size of the objects depicted in the image (again, another concern for Vishwanath's account). So instead of claiming that:

> The subjects varied their reach into pictorial space by 7 cm *because* they experienced stereoscopic depth (the 'coulisses' effect).

We can just as plausibly claim that:

> In the absence of strong stereoscopic information, subjects estimated the distances in the scene on the basis of the available pictorial cues.

Nor should it be claimed that reaching is somehow immune from cognitive influences in a way that verbal reports are not; as Firestone and Scholl (2016) observe, 'that seems absurd: Our object-directed actions can and do incorporate what we think, know, and judge—in addition to what we see…'.

iii. Visual Comparison: The only solution then is to test stereopsis in the monocular case in exactly the same way that we test stereopsis in the binocular case, namely with a comparison of the depth between two points (on the use of point source comparisons in the binocular context, see Palmisano et al. 2010). But isn't there a problem here? As Koenderink et al. (2011) observe, when monocularly viewed comparators are introduced into an image (for instance, Koenderink et al.'s Tissot gauges, see Fig. 3a), they are treated as *pictorial* objects in *pictorial* space. If the depth they record is merely pictorial, this won't help us to establish whether pictorial depth is itself stereoscopic? So how can we ensure that the depth that visual comparators record is *stereoscopic*, rather than merely *pictorial*?

The solution is to view the 2D image synoptically (i.e. present it individually to both eyes in a stereoscope), whilst testing the comparative depth of various points in the image using stereoscopic point sources. In this way, the impression of monocular depth can be preserved, whilst ensuring that the binocular depth of the point sources and the synoptic depth of the image are *directly* comparable. But what is interesting is that even though Koenderink et al. often employ synoptic viewing conditions they steadfastly refuse to adopt this approach. For instance, in Doorschot et al. (2001), they insist that: 'The gauge figure was presented monocularly to prevent the subjects from matching local disparities of the object and the gauge figure'. But if the monocular depth impressions are genuinely stereoscopic, then why should this matter?

As Bernhard et al. (2016) have recently demonstrated, estimates of the scene geometry of unfamiliar pictorial scenes are *vastly reduced* when evaluated using a *stereoscopic* Tissot gauge under synoptic conditions, which leads us to seriously question whether Koenderink et al.'s *monocular* Tissot gauge is really tracking stereoscopic depth. Indeed, Doorschot et al. (2001) begin to wonder whether their results really reflect our visual experience given that their monocular estimates of shape are so close to their binocular estimates: 'Given the immediate impression of the stimuli, it is perhaps surprising that we did not find a larger scaling effect (e.g. ... compare the impression of monocularly viewing one photo of a stereo pair to the binocularly fused impression)'.

Whilst it is true that *some* depth was still reported by Bernhard et al.'s using a *stereoscopic* Tissot gauge, this can be explained by the fact that *stereoscopic* Tissot gauges still contain a strong pictorial cue to slant, thereby affording a strong *cognitive* cue to depth; indeed, the fact that depth increased when images of familiar objects were used only confirms the significant *cognitive* dimension to this matching task. But we can largely eradicate this *cognitive* dimension if, instead of using a Tissot gauge to estimate local slant, we use two stereoscopic point sources to measure the depth between various points in the image. Indeed, measuring the depth between various points in the image is a test that Koenderink et al. (2011, 2015a) and Volcic et al. (2014) already employ (see Fig. 5), so we are simply ensuring that these studies reflect our *perception* rather than the *cognitive* influences inherent in our *judgement* (Koenderink et al.) or *action* (Volcic et al.).

iv. Comparison with Real Objects: The Japanese state broadcaster (NHK) has suggested that when 2D images are viewed monocularly on their 8K Super Hi-Vision displays, they are virtually indistinguishable from the real thing. Masaoka et al. (2013) showed subjects two stimuli in succession, a 2D image and a real object, and asked them to pick the one 'that appeared most similar to the real object'. Although subjects were marginally more likely to choose the real object, this result did not reach statistical significance.

But the force of this study depends on the conditions under which the stimuli were presented, and here I have a number of concerns: First, in order to control for binocular disparity, Masaoka et al. ensured that the stimuli were viewed under synoptic conditions. They recognise that this might *increase* the depth of the image, but they overlook the fact that it is also liable to *flatten* the real object itself, producing 'an uncanny flat appearance of a kind one cannot obtain by just closing one eye' (Koenderink et al. 1994); indeed, the subjects in Masaoka et al. rated the real objects as being *less than fully real*. Second, these 50 cm objects were viewed from a distance of 4.8 m, which we might expect would obviate any monocular stereoscopic impression. Third, Masaoka et al. admit that their experiment was aided by a narrow field of view and dark conditions. Finally, there was a significant memory component: The first stimulus was shown for 10 s, then there was a 4 s interval, then the second stimulus was shown for 10 s, and only then did the subjects make their choice.

But is there an alternative? Well, we might begin by (a) viewing a 2D image under monocular rather than synoptic conditions, and then (b) introducing a real object by means of a Pepper's Ghost effect: Does the real object simply fill the space *already* taken up by the monocular stereopsis from the 2D image? Or does the emergence of the real object *transform* the nature of the depth we see?

Finally, this is not to dismiss the findings in Masaoka et al. (2013). Concerned that the synoptic viewing conditions might explain their results, Masaoka et al. respond that even in normal viewing conditions 'images look nicely rounded and solid, which implies that higher angular resolution as well as synoptic viewing enhances pictorial relief'. But given that this observation occurred binocularly, and presumably at close quarters, this sounds less like stereopsis and more like the ability of pictorial cues to *camouflage* the flatness of a display. Interestingly, Vishwanath (2010) also articulates the inability to attribute flatness to 2D images in terms of camouflage: 'The ubiquity of camouflage suggests that animals do not see reflectance changes as such...'. And both Vishwanath and I agree that the inability to attribute flatness to 2D images need not imply the presence of positive stereoscopic depth. Instead, our only disagreement is as to whether this inability to attribute flatness to pictures is *perceptual* or merely *cognitive*? Indeed, this debate extends beyond pictorial depth to Zoology where the sharp distinction between cryptic-patterning as *perceptual* camouflage and mimicry as *cognitive* camouflage is only just beginning to erode (see Skelhorn and Rowe 2016).

REFERENCES

Akeley, K., Watt, S. J., Girshick, A. R., & Banks, M. S. (2004). A stereo display prototype with multiple focal distances. *ACM Transactions on Graphics, 23*(3), 804–813.

Ames, A., Jr. (1925a). The illusion of depth from single pictures. *Journal of the Optical Society of America, 10*(2), 137–148.

Ames, A., Jr. (1925b). Depth in pictorial art. *The Art Bulletin, 8*(1), 4–24.

Austin, J. L. (1962). *Sense and sensibilia*. Oxford: Oxford University Press.

Auvray, M., & O'Regan, J. K. (2003). Influence of semantic factors on blindness to progressive changes in visual scenes / L'influence des facteurs sémantiques sur la cécité aux changements progressifs dans les scènes visuelles. *L'année Psychologique, 103*, 9–32.

Balzer, R. (1998). *Peepshow: A visual history*. New York: Harry N. Abrams.

Barry, S. (2009). *Fixing my gaze: A scientist's journey into seeing in three dimensions*. New York: Basic Books.

Bayne, T., & Montague, M. (2011). *Cognitive phenomenology*. Oxford: Oxford University Press.

Bernhard, M., Waldner, M., Plank, P., Soltészová, V., & Viola, I. (2016). The accuracy of gauge-figure tasks in monoscopic and stereo displays. *IEEE Computer Graphics and Applications, 36*(4), 56–66.

Blake, E. C. (2003). Zograscopes, virtual reality, and the mapping of polite society in eighteenth-century England. In L. Gitelman & G. B. Pingree (Eds.), *New media* (pp. 1–30). Cambridge, MA: MIT Press.

Bradshaw, M. F., Parton, A. D., & Glennerster, A. (2000). The task-dependent use of binocular disparity and motion parallax information. *Vision Research, 40*(27), 3725–3734.

Carr, H. A. (1935). *An introduction to space perception*. New York: Longmans, Green, & Co.

Chudnoff, E. (2015). *Cognitive phenomenology*. Oxford: Routledge.

Claparède, E. (1904). Stéréoscopie monoculaire paradoxale. *Annales d'Oculistique, 132*, 465–466.

Cooper, E. A., & Banks, M. S. (2012). Perception of depth in pictures when viewed from the wrong distance. *Journal of Vision, 12*(9), 896 (abstract).

Cutting, J. E. (2003). Reconceiving perceptual space. In H. Hecht, R. Schwartz, & M. Atherton (Eds.),*Perceiving Pictures: An Interdisciplinary Approach to Pictorial Space*. Cambridge, MA: MIT Press.

Di Luca, M., Domini, F., & Caudek, C. (2010). Inconsistency of perceived 3D shape. *Vision Research, 50*(16), 1519–1531.

Dolgoff, E. (1997). Real-depth imagining. *SID Digest, 28*, 269–272.

Domini, F., & Caudek, C. (2011). Combining image signals before three-dimensional reconstruction: The intrinsic constraint model of cue integration. In Trommershäuser, Körding, & Landy (Eds.), *Sensory cue integration*. Oxford: Oxford University Press.

Doorschot, P. C., Kappers, A. M., & Koenderink, J. J. (2001). The combined influence of binocular disparity and shading on pictorial shape. *Perception and Psychophysics, 63*, 1038–1047.

Elner, K. W., & Wright, H. (2015). Phenomenal regression to the real object in physical and virtual worlds. *Virtual Reality, 19*(1), 21–31.

van Ee, R, van Dam, L. C. J., & Erkelens, C. J. (2002). Bi-stability in perceived slant when binocular disparity and monocular perspective specify different slants. *Journal of Vision, 2*(9), 597–607.

Erkelens, C. J. (2012). Contribution of disparity to the perception of 3D shape as revealed by bistability of stereoscopic Necker cubes. *Seeing and Perceiving, 25*(5), 561–576.

Erkelens, C. J. (2013). Virtual slant explains perceived slant, distortion and motion in pictorial scenes. *Perception, 42*, 253–270.

Fine, I., Wade, A. R., Brewer, A. A., May, M. G., Goodman, D. F., Boynton, G. M., et al. (2003). Long-term deprivation affects visual perception and cortex. *Nature Neuroscience, 6*(9), 915–916.

Firestone, C., & Scholl, B. J. (2016). Seeing and thinking: Foundational issues and empirical horizons. *Behavioral and Brain Sciences, 39*, 53–67.

Gabor, D. (1960, July 14). Three-dimensional cinema. *New Scientist, 141*.

Gibson, J. J. (1947). *Motion picture testing and research*. Research Reports, Report No. 7, Army Air Forces Aviation Psychology Program.

Gibson, J. J. (1966). *The senses considered as perceptual systems*. Boston: Houghton Mifflin.

Glennerster, A., Rogers, B. J., & Bradshaw, M. F. (1996). Stereoscopic depth constancy depends on the subject's task. *Vision Research, 36*(21), 3441–3456.

Gogel, W. C. (1969). The sensing of retinal size. *Vision Research, 9*, 3–24.

Gogel, W. C., & Da Silva, J. A. (1987). Familiar size and the theory of off-sized perceptions. *Perception and Psychophysics, 41*(4), 318–328.

Goodale, M. A., & Servos, P. (1996). Visual control of prehension. In H. Zelaznik (Ed.), *Advances in motor learning and control* (pp. 87–121). Champaign, IL: Human Kinetics Publishers.

Gregory, R. L. (1968). Perceptual illusions and brain models. *Proceedings of the Royal Society B, 171,* 279–296.

Gregory, R. L. (2004). The blind leading the sighted. *Nature, 430,* 1.

Gregory, R. L. (2005). The Medawar lecture 2001 knowledge for vision: Vision for knowledge. *Philosophical Transactions of the Royal Society B, 360,* 1231–1251.

Gregory, R. L., & Wallace, J. G. (1963). *Recovery from early blindness: A case study.* Experimental Psychology Society Monograph, no. 2.

Hagen, M. A. (1980). *The Perception of Pictures I: Alberti's Window: The Projective Model of Pictures.* New York: Academic Press.

Hibbard, P. (2008). Can appearance be so deceptive? Representationalism and binocular vision. *Spatial Vision, 21*(6), 549–559.

Hill, H., & Bruce, V. (1993). Independent effects of lighting, orientation, and stereopsis on the hollow-face illusion. *Perception, 22,* 887–897.

Holt, E. (1904). Die von M. von Rohr gegebene Theorie des Veranten, eines Apparats zur Richtigen Betrachtung von Photographien by E. Wandersleb; The Verant, a New Instrument for Viewing Photographs from the Correct Standpoint by M. von Rohr; Der Verant, ein Apparat zum Betrachten von Photogrammen in Richtigen Abstande by A. Köhler. *The Journal of Philosophy, Psychology and Scientific Methods, 1*(20), 552–553.

Hornsey, R. L., Hibbard, P. B., & Scarfe, P. (2015). Ordinal judgments of depth in monocularly- and stereoscopically-viewed photographs of complex natural scenes. *Proceedings of the International Conference on 3D Imaging.*

Howard, I. P., & Rogers, B. J. (2012). *Perceiving in depth.* Oxford: Oxford University Press.

Hyde, R. (2015). *Paper peepshows: The Jaqueline and Jonathan Gestetner collection.* Suffolk: Antique Collectors' Club.

Johnston, E. B. (1991). Systematic distortions of shape from stereopsis. *Vision Research, 31*(7–8), 1351–1360.

Johnston, E. B., Cumming, B. G., & Parker, A. J. (1993). Integration of depth modules: Stereo and texture. *Vision Research, 33,* 813–882.

Julesz, B. (1960). Binocular depth perception of computer-generated patterns. *Bell Labs Technical Journal, 39,* 1125–1162.

Knill, D. C., & Saunders, J. A. (2003). Do humans optimally integrate stereo and texture information for judgments of surface slant? *Vision Research, 43*(24), 2539–2558.

Koenderink, J. J. (1998). Pictorial relief. *Philosophical Transactions of the Royal Society A, 356,* 1071–1086.

Koenderink, J. J. (2011). Gestalts and pictorial worlds. *Gestalt Theory, 33,* 289–324.

Koenderink, J. J. (2015). PPP. *Perception, 44,* 473–476.

Koenderink, J. J., & van Doorn, A. J. (1995). Relief: Pictorial and otherwise. *Image and Vision Computing, 13,* 321–334.

Koenderink, J. J., & van Doorn, A. J. (2003). Pictorial space. In H. Hecht, R. Schwartz, & M. Atherton (Eds.), *Perceiving Pictures: An Interdisciplinary Approach to Pictorial Space.* Cambridge, MA: MIT Press.

Koenderink, J. J., van Doorn, A. J., & Kappers, A. M. L. (1994). On so-called paradoxical monocular stereoscopy. *Perception, 23,* 583–594.

Koenderink, J. J., van Doorn, A. J., & Kappers, A. M. L. (2006). Pictorial relief. In M. R. M. Jenkin & L. R. Harris (Eds.), *Seeing spatial form.* Oxford: Oxford University Press.

Koenderink, J. J., van Doorn, A., & Wagemans, J. (2011). Depth. *i-Perception, 2,* 541–564.

Koenderink, J., van Doorn, A., & Wagemans, J. (2015a). Deploying the mental eye. *i-Perception, 6*(5), 1–17.

Koenderink, J., van Doorn, A., & Wagemans, J. (2015b). Part and Whole in Pictorial Relief. i-Perception, 6(6), 1–21.

Koenderink, J. J., van Doorn, A. J., Kappers, A. M. L., & Todd, J. T. (2004). Pointing out of the picture. *Perception, 33,* 513–530.

Koenderink, J., van Doorn, A., Albertazzi, L., & Wagemans, J. (2015c). Relief articulation techniques. *Art & Perception, 3*(2), 151–171.

Koenderink, J. J., Wijntjes, M. W. A., & van Doorn, A. J. (2013). Zograscopic viewing. *i-Perception, 4*(3), 192–206.

Koenderink, J. J., et al. (2010). Does monocular visual space contain planes? *Acta Psychologica, 134,* 40–47.

Kubovy, M. (1986). *The psychology of perspective and renaissance art.* Cambridge: Cambridge University Press.

Landy, M., Banks, M., & Knill, D. (2011). Ideal-observer models of cue integration. In Trommershäuser, Körding, & Landy (Eds.), *Sensory cue integration.* Oxford: Oxford University Press.

Liu, S., Cheng, D. W., & Hua, H. (2008). An optical see-through head-mounted display with addressable focal planes. In *Proceedings of IEEE/ACM Intl Symp. Mixed and Augmented Reality (ISMAR 08),* 33–42.

Loftus, A., Servos, P., Goodale, M., Mendarozqueta, N., & Mon-Williams, M. (2004). When two eyes are better than one in prehension: Monocular viewing and end-point variance. *Experimental Brain Research, 158*(3), 317–327.

Loomis, J. M., Philbeck, J. W., & Zahorik, P. (2002). Dissociation between location and shape in visual space. *Journal of Experimental Psychology: Human Perception and Performance, 28*(5), 1202–1212.

Love, G. D., Hoffman, D. M., Hands, P. J. W., Gao, J., Kirby, A. K., & Banks, M. S. (2009). High-speed switchable lens enables the development of a volumetric stereoscopic display. *Optics Express, 17*(18), 15716–15725.

Masaoka, K., Nishida, Y., Sugawara, M., Nakasu, E., & Nojiri, Y. (2013). Sensation of realness from high-resolution images of real objects. *IEEE Transactions on Broadcasting, 59*, 72–83.

Mausfeld, R. (2003). Conjoint Representations and the Mental Capacity for Multiple Simultaneous Perspectives. In H. Hecht, R. Schwartz, & M. Atherton (Eds.), *Perceiving Pictures: An Interdisciplinary Approach to Pictorial Space.* Cambridge, MA: MIT Press.

Metzger, W. (1975). *Gesetze des Sehens.* Frankfurt: Waldemar Kramer.

Münsterberg, H. (1904). Perception of distance. *Journal of Philosophy, Psychology and Scientific Methods, 1*(23), 617–623.

Ooi, L. T., & He, Z. J. (2015). Space perception of strabismic observers in the real world environment. *Investigative Ophthalmology & Visual Science, 56*, 1761–1768.

O'Regan, J. K., & Noë, A. (2001). A sensorimotor account of vision and visual consciousness. *Behavioral and Brain Sciences, 24*(5), 883–917.

O'Regan, J. K. (2011). *Why Red Doesn't Sound Like a Bell: Understanding the Feel of Consciousness.* Oxford: Oxford University Press.

Palmisano, S., Gillam, B., Govan, D. G., Allison, R. S., & Harris, J. M. (2010). Stereoscopic perception of real depths at large distances. *Journal of Vision, 10*(6), 19.

Papathomas, T. V. (2002). Experiments on the role of painted cues in Hughes's reverspectives. *Perception, 31*, 521–530.

Parker, A. J. (2016). Vision in our three-dimensional world. *Philosophical Transactions of the Royal Society B, 371*(1697), 20150251.

Peacocke, C. (1983). *Sense and content: Experience, thought, and their relations.* Oxford: Oxford University Press.

Predebon, J. (1993). The familiar-size cue to distance and stereoscopic depth perception. *Perception, 22*(8), 985–295.

Ramachandran, V. S. (1988). Perception of shape from shading. *Nature, 331*, 163–166.

Rogers, B. J., & Bradshaw, M. F. (1993). Vertical disparities, differential perspective and binocular stereopsis. *Nature, 361*, 253–255.

Rogers, B. J., & Bradshaw, M. F. (1995). Disparity scaling and the perception of frontoparallel surfaces. *Perception, 24*, 155–179.

Rogers, B., & Gyani, A. (2010). Binocular disparities, motion parallax, and geometric perspective in Patrick Hughes's 'reverspectives': Theoretical analysis and empirical findings. *Perception, 39*(3), 330–348.

Rolland, J. P., Krueger, M. W., & Goon, A. (2000). Multi-focal planes head-mounted displays. *Applied Optics, 39*(19), 3209–3215.

Sacks, O. (2006). Stereo Sue. In O. Sacks (Ed.), *The minds eye (2010)*. New York: Picador.

Sacks, O. (2010). Persistence of vision: A journal. In O. Sacks (Ed.), *The minds eye (2010)*. New York: Picador.

Scarfe, P., & Hibbard, P. B. (2013). Reverse correlation reveals how observers sample visual information when estimating three-dimensional shape. *Vision Research, 86*, 115–127.

Schlosberg, H. (1941). Stereoscopic depth from single pictures. *The American Journal of Psychology, 54*(4), 601–605.

Schwitzgebel, E. (2006). Do things look flat? In E. Schwitzgebel (Ed.), *Perplexities of consciousness (2011)*. Cambridge, MA: MIT Press.

Servos, P. (2000). Distance estimation in the visual and visuomotor systems. *Experimental Brain Research, 130*, 35–47.

Servos, P., & Goodale, M. A. (1994). Binocular vision and the on-line control of human prehension. *Experimental Brain Research, 98*, 119–127.

Servos, P., Goodale, M. A., & Jakobson, L. S. (1992). The role of binocular vision in prehension: A kinematic analysis. *Vision Research, 32*, 1513–1521.

Simons, D. J., Franconeri, S. L., & Reimer, R. L. (2000). Change blindness in the absence of a visual disruption. *Perception, 29*(10), 1143–1154.

Skelhorn, J., & Rowe, C. (2016). Cognition and the evolution of camouflage. *Proceedings of the Royal Society B, 283*(1825), 20152890.

Smithies, D. (2013). The nature of cognitive phenomenology. *Philosophy Compass, 8*(8), 744–754.

Strawson, G. (1994). *Mental reality*. Cambridge, MA: MIT Press.

Thouless, R. H. (1931a). Regression to the real object I. *British Journal of Psychology, 21*(4), 339–359.

Thouless, R. H. (1931b). Regression to the real object II. *British Journal of Psychology, 22*(1), 1–30.

Todd, J. T., & Norman, J. F. (2003). The visual perception of 3-D shape from multiple cues: Are observers capable of perceiving metric structure? *Perception and Psychophysics, 65*, 31–47.

Travis, C. S. (2004). The silence of the senses. *Mind, 113*(449), 57–94.

Travis, C. S. (2013). The silence of the senses. In *Perception: Essays after Frege*. Oxford: Oxford University Press.

Tye, M. (1993). Blindsight, the absent qualia hypothesis, and the mystery of consciousness. *Royal Institute of Philosophy Supplement, 34*, 19–40.

Tye, M. (2003). Consciousness, color, and content. *Philosophical Studies, 113*(3), 233–235.

Vishwanath, D. (2010). Visual information in surface and depth perception: Reconciling pictures and reality. In Albertazzi, van Tonder, & Vishwanath (Eds.), *Perception beyond inference: The informational content of visual processes*. Cambridge, MA: MIT Press.

Vishwanath, D. (2013). Experimental phenomenology of visual 3d space: Considerations from evolution, perception, and philosophy. In L. Albertazzi (Ed.), *Handbook of experimental phenomenology*. Chichester: Wiley-Blackwell.

Vishwanath, D. (2014). Towards a new theory of stereopsis. *Psychological Review, 121*(2), 151–178.

Vishwanath, D. (2016). Induction of monocular stereopsis by altering focus distance: A test of Ames's hypothesis. *i-Perception, 7*(2), 1–5.

Vishwanath, D., & Blaser, E. (2010). Retinal blur and the perception of egocentric distance. *Journal of Vision, 10*(10), 26.

Vishwanath, D., & Domini, F. (2013). Pictorial depth is not statistically optimal. *Journal of Vision, 13*(9), 613 (abstract).

Vishwanath, D., & Hibbard, P. B. (2013). Seeing in 3D with just one eye: Stereopsis without binocular vision. *Psychological Science, 24*(9), 1673–1685.

Volcic, R., Vishwanath, D., & Domini, F. (2014). Reaching into pictorial spaces. In *Proceedings of SPIE, 9014: Human Vision and Electronic Imaging XIX.*

von Hildebrand, A. (1893). *Das Problem der Form in der bildenden Kunst.* Strasbourg: Heitz.

von Rohr, M. (1903). The verant, a new instrument for viewing photographs from the correct standpoint. *The Photographic Journal, 43*, 279–290.

Watt, S. J., Akeley, K., Girshick, A. R., & Banks, M. S. (2005). Achieving near-correct focus cues in a 3-D display using multiple image planes. In *Proceedings of SPIE: Human Vision and Electronic Imaging, (IS&T/SPIE Paper Number 5666-53).*

Wijntjes, M. W. A., & Pont, S. C. (2012). Perceived depth in photographs: Humans perform close to veridical on a relative size task. *Journal of Vision, 12*(9), 277 (abstract).

Wijntjes, M. W. A., Füzy, A., Verheij, M. E. S., Deetman, T., & Pont, S. C. (2016). The synoptic art experience. *Art & Perception, 4*(1–2), 73–105.

CHAPTER 4

The Physiology and Optics of Monocular Stereopsis

Abstract Whilst Gibson (1947) gives us an explanation of depth perception for the monocular observer in motion, and Julesz (1960) explains depth perception for the static binocular observer, Physiological Optics has had less success explaining depth perception for the static monocular observer. In this chapter I attempt to outline what a Physiological Optical account of the static monocular observer might look like: First, I argue that we should be sceptical of treating accommodation as a depth cue and should instead be open to the idea that monocular vision may be unable to convey a sense of scale. Second, I suggest that the phenomenal geometry of the scene might be provided by defocus blur, but only once it has been appropriately reconceived.

Keywords Accommodation · Vergence · Defocus blur Chromatic aberration · Magnification

In the first section of this chapter I explore whether the visual system has access to an egocentric distance cue by which to scale the visual geometry of the scene. In terms of monocular distance cues, accommodation (as measured by the tension in the ciliary muscles required to bring an object into

---

The original version of this chapter was revised: Post-publication corrections have been incorporated. The erratum to this chapter is available at https://doi.org/10.1007/978-3-319-66293-0_5

© The Author(s) 2017
P. Linton, *The Perception and Cognition of Visual Space*,
DOI 10.1007/978-3-319-66293-0_4

focus) is an obvious candidate. But the more we explore accommodation as a potential distance cue, the less it seems to perform this function. So perhaps scale is something we only cognitively impute to the monocular scene? In the second section, I explore whether defocus blur functions as a cue to the visual geometry of the scene and explain why we must conceive of defocus blur as an optical cue rather than a pictorial cue if it is to perform this function.

1   SCALE

One of the most basic questions in depth perception is whether vision conveys an impression of scale? For instance, does a small cardboard cut-out viewed up close really look different from a large cardboard cut-out viewed from far away? Or is scale something that we merely cognitively impute to the visual scene? For instance, Helmholtz (1866) regarded familiar size as an important cue, but the last 50 years has seen a realisation that familiar size does not appear to affect the distance at which a familiar object is *seen*, as opposed to *judged*. As Gogel (1969) explains:

> The distance reports made under these conditions probably should be considered to be inferential rather than perceptual, i.e. more determined by the tendency to report that an object is distant because its perceived size is small than to report the distance actually perceived.

But as I discuss below, I do not think that Gogel's own preferred solution, namely, the specific distance tendency, is any more perceptual: The fact that an object with no cues to size or distance will be judged to be 4–12 ft from the observer is equally open to the criticism that it is merely a best guess in a situation of uncertainty rather than true perception (Howard and Rogers 2012). Nor is 4–12 ft particularly specific for a specific distance tendency, which only confirms the impression that it is the best guess.

1.1   Accommodation: Physiological or Optical Cue?

More promising is accommodation. In *optics* accommodation refers to the focal distance of the eye, but in the depth perception literature it is generally measured in *physiological* terms, namely the degree of tension in the ciliary muscles required to bring an object into focus: The closer the object, the greater the tension. And it is often supposed that once the visual system has specified the distance of the focal plane in this way, this information could be used to scale the scene as a whole.

But even if the visual system could specify the distance of the focal plane, it is not at all clear that a *physiological* mechanism would be responsible. For instance, it might be the case that distance is specified *optically* in the retinal image itself: Since the intraocular lens is not optically perfect, but subject to a number of aberrations, the degree of contraction in the intraocular lens might be left as a *trace* or *signature* on the retinal image (so long as these aberrations vary predictably with the power of the lens, which many of them appear to, for instance chromatic aberration, spherical aberration, and, although not strictly speaking an aberration, low-frequency microfluctuations). In which case, these *aberrations* would not to be seen as *defects*, but as valuable sources of distance information.

This suggestion relates to a broader tension in the literature: On the one hand, there is the temptation to conduct vision science at the level of *ideal theory*. For instance, in 2007–2009, the editorial pages of *Perception* played host to a debate between the two leading conceptions of visual optics: the Optic Array (defended by Gillam 2007, 2009 and Rogers 2007, 2009) and the Retinal Image (defended by Frisby 2009 and Gregory 2009). But what was so startling about this debate was how quick both sides were to disavow anything other than *idealised* optics: For instance, Rogers (2007) insists that the very appeal of the optic array lies in the fact that: 'Eyes do not come into an optic array description', and so leaves visual optics 'uncontaminated by the distortions, imperfections, and philosophical confusions that belie our retinal images'. Whilst Frisby (2009) sought to defuse Rogers' concerns by shifting our attention away from the *real* retinal images of Physiological Optics, and towards the *idealised* retinal image of computer vision.

But at the same time, another debate was occurring in the pages of the *Journal of Optometry* between Navarro (2009a), Kruger (2009), and Navarro (2009b). As Kruger concludes, in his response to Navarro:

> While our idealized model of the eye remains constrained by zeitgeist and dogma, Darwin's eye, less encumbered by word or name or myth, embraces nature's 'crude flaws' as distance and relative depth...

In terms of *relative depth*, Kruger cites Nguyen et al. (2005), which will be discussed in Sect. 2.2. In terms of *distance*, Kruger appears to be referring a point he makes earlier in the article, namely that accommodation can respond without feedback from defocus blur (citing Kruger et al. 1997), which implies that something else must be determining the *magnitude* of the accommodative response. Whether this is the correct interpretation is open to debate: As Morrison et al. (2010) observe, open-loop accommodation is heavily determined by the method employed to open the feedback loop, and Kruger et al. (1997) employ a 3 mm pupil (vs. the 0.33 mm pupil that Morrison et al. 2010 identify as the gold standard today). In any case, Kruger rejects chromatic aberration and higher order monochromatic aberrations (such as spherical aberration) as potential sources for this information, suggesting that a Stiles–Crawford-esque 'sensitivity to the wavefront vergence across the exit pupil of the eye' may determine the magnitude of the response.

But the debate between Navarro and Kruger opens up an intriguing possibility: Could the visual system use optical aberrations to extract the scale of the scene? You might think this question has already been considered; for instance, Held et al. (2010) and Vishwanath and Blaser (2010) document the apparent miniaturisation that comes with depth of field blur.

In Fig. 1b, Held et al. apply a linear blur gradient to the aerial image in Fig. 1a. This blur gradient is consistent with the original scene being photographed with a 60 m wide aperture, but this is not the interpretation that we automatically adopt. Instead, we appear to attribute the presence of a linear blur gradient to the optics of the eye itself; for instance, the gradient in Fig. 1b is consistent with viewing the scene from 6 cm. Consequently, Held et al. suggest that we experience the scene as a miniaturised model. I quite agree, but the question is whether the effect in Fig. 1b is *perceptual* or *cognitive*? Do we literally *see* the image in Fig. 1b as being closer, or do we merely *interpret* it as such?

A perceptual interpretation becomes hard to sustain when we view Fig. 1a, b alongside one another on the page: They seem to be at the very same distance. Furthermore, the elements in Fig. 1b don't appear to take up any less of the visual field than the objects in Fig. 1a. So in what sense has there been a *visual* miniaturisation? A better interpretation, and one more in keeping with my account of pictorial cues advanced in Chap. 3, is that this miniaturisation effect occurs purely at the level of cognition; indeed, we might draw an analogy between the purely cognitive imputation of scale with *familiar size* that Gogel (1969)

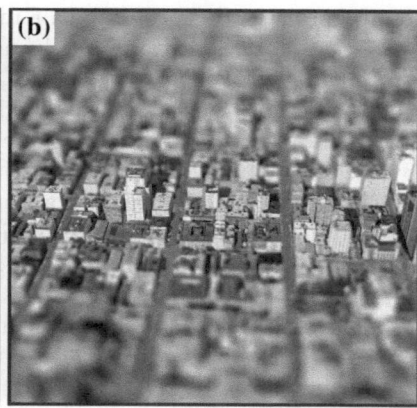

Fig. 1   Held et al. (2010) rendering of a cityscape using **a** a pinhole aperture and **b** a 60 m aperture. From Held et al. (2010). Using blur to affect perceived distance and size. ACM Transactions on Graphics, 29(2): 19. © Association for Computing Machinery, Inc. Original city images and data from GoogleEarth are copyright Terrametrics, SanBorn, and Google

investigated, and the purely cognitive imputation of scale with *familiar blur* that Held et al. and Vishwanath and Blaser investigate.

By contrast, Vishwanath and Blaser argue that the miniaturisation apparent in Fig. 1b must be *visual* because it doesn't just bias subjects' estimates of the *distance depicted* in the images, but it also biases subjects' estimates of the *physical distance* of the display itself. But one of the recurring questions of Chaps. 2 and 3 is whether such biases are (a) *perceptual* (as Vishwanath & Blaser contend), (b) reflect a *conscious decision strategy*, or (c) are the result of an *unconscious post-perceptual judgement* that attributes meaning to what we see? For instance, a bias along the lines of (c) could clearly affect the *evaluative judgements* relied upon in Vishwanath and Blaser (ordinal depth between two displays viewed in succession). Instead, a truly *perceptual* test would be to view Fig. 1b synoptically, set a frame at the distance of the display (which the subject never loses sight of), and then replace Fig. 1b with Fig. 1a: Unless Fig. 1a appears *behind* the frame, or the frame appears to *move backwards in space* when Fig. 1a is shown, it would be hard to maintain that the miniaturisation effect in Fig. 1b is really perceptual rather than cognitive.

But perhaps the possibility of intraocular lens aberrations leaving a *trace* or *signature* on the retinal image does open up the prospect of a genuinely *optical* cue to scale? I fear, however, that we are left in a rather unenviable position: We know the visual system has access to this *optical* information on the scale but it doesn't appear to use it. And there are two reasons for coming to this conclusion.

The first is that controlling for optical aberrations appears to have relatively little effect on subjects' abilities to either (a) *accommodate* (or, at the very least, *accommodate in the right direction*) or (b) *judge ordinal depth*, both of which you would expect optical aberrations would contribute to if they were effective depth cues. In the context of (a) *accommodation in the right direction*, this has been demonstrated in the context of (i) astigmatism (second-order monochromatic aberration), see Gambra et al. (2009), even though, as Burge and Geisler (2011) observe, 'astigmatism is deliberately added to the lenses in compact disc players to aid their autofocus devices' (see Cope 1993); (ii) coma and trefoil (third-order monochromatic aberrations), see Fernández and Artal (2005) and López-Gil et al. (2007), as well as the fact that at least in theory, *myopic* and *hyperopic* blur ought to produce identical retinal images for odd-order aberrations; and (iii) spherical aberration (fourth-order monochromatic aberration), see Chen et al. (2006); Chin et al. (2009b); Hampson et al. (2010); Wahlberg et al. (2011), cf. Chin et al. (2009a). And, in the context of (b) *ordinal depth judgements*: (iv) low-frequency microfluctuations, see Zannoli et al. (2016), and (v) chromatic aberrations, see Nguyen et al. (2005) and Zannoli et al. (2016), although my reading of Nguyen et al. is controversial and will have to be justified in Sect. 2.2.

Now, the ability to accommodate in the right direction when aberrations have been disrupted isn't fatal to the argument that optical aberrations might contribute information about *scale*; after all, with defocus blur appearing to control the *magnitude* of the accommodative response, all that the visual system needs from optical aberrations in order to shift accommodation from one object to another is the *direction* of that second object and this might come from other sources, see for instance López-Gil et al. (2016) on the presence of blood vessels in front of the photoreceptor cells of the retina (which seems a more promising candidate than the Stiles–Crawford effect, see Kruger et al. 2001, 2004; Stark et al. 2009). So perhaps the enumerated optical aberrations might be essential to depth perception even though they are not essential to accommodation?

But I would be sceptical of putting too much emphasis on the distinction between the information required for accommodation and the information required for depth perception. It is true that there is a tendency to think of the relationship between optical aberrations, accommodation, and depth perception, in terms of:

[A] Optical Aberrations → Accommodative Response → Perception of Distance

Or, at the very most (in light of Nguyen et al. 2005):

[B] Optical Aberrations → Accommodative Response → Perception of Distance

And independently:

Optical Aberrations → Disambiguation of Defocus Blur for Ordinal Depth

But there is increasing evidence that the link between optical aberrations and accommodation is mediated by *cognition*. As Gwiazda et al. (1993) observe, the accommodative response reflects more than just the optics of the scene, with the presence of proximity cues improving the accommodative response. Indeed, as Charman and Heron (2015) observe in the context of microfluctuations, attempting to isolate the accommodative reflex from other (supposedly confounding) cognitive influences leaves us writing off huge swathes of normal subjects:

> Often subjects fail to respond at all, particularly in studies involving Badal stimuli but also those where free-space viewing is involved. ... It is difficult to explain the behaviour of these subjects by any microfluctuation-based, simple 'reflex' model of accommodation.

For instance, only 4 out of the 25 subjects in Metlapally et al. (2014) were able to effectively respond to Badal stimuli (by contrast, most 'had sluggish or no reflex accommodation responses'), similarly only 5 out of the 25 subjects in Metlapally et al. (2016) were able to respond. So even as a *reflex*, accommodation would appear to have a *cognitive* dimension. Indeed, there appear to be echoes here of Holmes's (1918) and Holmes and Horrax's (1919) observation that brain damaged soldiers who lost spatial cognition also lost their blink reflex. Consistent with this conclusion, Otero Molins et al. (2016) and Aldaba et al. (2016) demonstrate the importance of pictorial cues for inducing saccades under Badal conditions, whilst Morrison et al. (2010) document the extent to which cognition has been known to influence 'open-loop' accommodation.

But it seems unlikely that optical aberrations would feed into our *cognition* of depth without first feeding into our *perception*: First because there doesn't appear to be a shortcut for these aberrations to inform *cognition* without having to go via *perception*. And second, think about our experience of shifting our gaze from one object to another: It's not as if the object we are about to shift to suddenly pops into depth just as we are about to make an eye movement. Instead, the object is *already* perceived as being in depth off the focal plane, and it is on the basis of this *perceptual* information that we ordinarily plan to shift our gaze from one object to another.

But these observations would appear to suggest the following relationship between optical aberrations and the accommodative response in most, and potentially even all, changes in monocular accommodation:

[C] Optical Aberrations → Depth Perception → Depth Cognition → Accommodation

So trying to distinguish too sharply between the optical information required for depth perception and the optical information required for accommodation may be a mistake.

1.2   *The Role of Accommodation and Vergence in Optometry*

The second reason I am sceptical that the visual system extracts the distance of the focal plane from optical aberrations is that I am sceptical that the visual system extracts the distance of the focal plane from any source whatsoever. This scepticism takes two forms: *theoretical* (in this section) and *empirical* (in the next section).

The source of the theoretical concern is this: Optometry has been a flourishing discipline since the thirteenth century, with hundreds of millions of people having enjoyed corrective lenses *without* distorting their visual space: optically correcting *myopia* and *hyperopia* adds clarity; it doesn't radically alter the scale of the scene. Indeed if it did, the cure would be worse than the disease. And yet if accommodation really is a depth cue then this is exactly what we ought to experience.

a. Myopia and Correction: To see why, consider the information that the brain receives when a myopic subject (of +6D) looks at an object 16 cm away. The visual system is completely relaxed, and the object is only *just* brought into focus: So *as far as the brain is concerned* the subject is looking at the horizon even though the object is only 16 cm away from the eye. This wouldn't be problematic if the subject *saw* the object as being located on the horizon but there is no evidence that they do.

But perhaps you might say that the visual system adapts to the myopia in this case, so relaxed accommodation is now seen as being located at 16 cm? But then we introduce a −6D lens to correct for the myopia, bringing an object that is on the horizon just into focus. But does the object located on the horizon appear to be located at 16 cm? No. Nor does the object located at 16 cm look as if its distance has been halved to 8 cm when it is brought into focus.

The concern that underpins these observations is this: For accommodation to be a depth cue, it necessarily presupposes that one of these 'settings' (corrected or uncorrected) *is the right one*, with the implication that other one must be *wrong*. By contrast, the visual experience of myopics appears to suggest that they have two equally permissible viewing modes: one with correction and one without; so long as the object is brought into focus it doesn't appear to matter. But if this is right, and *both* viewing modes are correct, then accommodation cannot be a depth cue.

So far the argument has been couched in terms of *misplacement*, i.e. does vision in one of these viewing modes *look wrong*? But it could just as easily be couched in terms of *difference*, i.e. does alternating between these two viewing modes *look significantly different* so far as distance is concerned? I say *significantly* different because *optical correction plus accommodation* is not a straight optical substitute for *uncorrected* vision since we are substituting intraocular lens accommodation for corneal accommodation and, because the intraocular lens lies some 4 mm behind the cornea, there may be a small change in magnification. Still, from 16 cm (6D) to the horizon (0D) is pretty much the full range of accommodation in adults and therefore the potential range of accommodation as a depth cue. So if +6D myopic subjects really don't notice a *significant* difference in depth as they focus on an object whilst a −6D contact lens is slid on and off the pupil, then it is hard to see how it could be maintained that accommodation is a depth cue.

b. Optometry versus Ophthalmology: Just as Optometry is a flourishing discipline, so too is Ophthalmology, and yet these two disciplines often take radically different approaches to correcting the very same optical complaint. But the fact that they both appear to be equally permissible appears to suggest that accommodation (and, dare I say, vergence) is not a depth cue. To put the point very crudely, Ophthalmology is concerned with achieving the right *physiological* outcome and therefore the right *optical* outcome, whilst Optometry is *only* concerned with achieving the right *optical* outcome irrespective of the *physiological* outcome. To be more specific, Ophthalmology is concerned with *repairing* whatever visual defect needs correcting and operates at the level of the visual system. By contrast, Optometry leaves the visual defect intact but merely *compensates* for it (or *counteracts* it) by operating at the level of the light that reaches the visual system: It is something we *add on to* the visual system rather than *do to* the visual system, an *external* rather than *internal* intervention.

But the fact that both of these forms of correction are equally permissible raises concerns for vergence and accommodation as depth cues.

Let me illustrate the point in the context of strabismus (subjects with misaligned eyes): Ophthalmology would remove the vergence error by either (a) performing surgery or (b) teaching the subject to control their own vergence movements. By contrast, Optometry *leaves the vergence error intact*. Instead, it simply asks subjects to wear prism glasses. In both cases, you get the right *optical* result (the right light hitting the retina), and therefore the right *visual* result (corrected vision), but in only one case do you achieve the right *physiological* result. But if this right, and we have two equally permissible ways of achieving the very same outcome, then this would appear to suggest that what matters is simply *getting the right optical result,* and so long as this is achieved the *physiology doesn't matter.* Indeed, this is quite vividly illustrated by the Optometric solution to strabismus: The visual deficit is that the eyes are crossed, and yet after prism lens 'correction', the subject's eyes *still* remain crossed; *all* that changes is the light directed to them.

The evidence in favour of vergence as a depth cue is significant (see Mon-Williams and Tresilian 1999; Viguier et al. 2001), but asking subjects to shift their gaze to a stationary stimulus brings with it a whole host of confounding cues (the initial diplopia in the stimulus, a visual impression of the process of vergence, and even at close distances a felt sensation in the ciliary muscles) and it is worth considering the extent to which these confounding cues might provide a cognitive basis for the performance in tasks such as Mon-Williams and Tresilian (1999), Viguier et al. (2001)? For instance, Morrison & Whiteside (1984) found that subjects are surprisingly good at judging distance on the basis of diplopia alone. Furthermore, whilst Tresilian et al. (1999) found that subjects could roughly judge the distance of a stimulus using vergence ($y = 0.39x + 36.27$), they found that subjects were twice as accurate if they simply relied on a monocular size cue ($y = 0.73x + 11.77$). Since the size of a nondescript stimulus should give us no information about its absolute distance, and since performance with vergence and the size cue together was only marginally better than the size cue alone ($y = 0.86x + 11.6$), we really have to question the extent to which the perception of distance is contributing to our distance estimates in vergence tasks.

But our primary concern in this chapter is with accommodation. Imagine a patient with instrument myopia: when they look down a microscope, they can't help but over-accommodate. And imagine we had to treat this patient. On the one hand, we could administer a cycloplegic drug to relax their ciliary muscles: akin to Ophthalmology, this would achieve the right *optical* outcome by achieving the right *physiological* outcome. On the other hand, we could simply achieve the right *optical* outcome by inserting a minus lens between the eye and the microscope, whilst leaving the physiological defect (the over-accommodation) intact. If this is right, and both solutions are equally permissible, then it would appear that *all* that matters is the right *optical* outcome, and *physiology* drops out of the equation.

1.3   Testing Accommodation as a Depth Cue

But how would we go about testing this hypothesis? Koenderink et al. (2011) appeal to the fact that distance is often *indeterminate* in the absence of contextual cues:

As one looks into a Ganzfeld [luminous surface] the awareness is that of a luminous foggy atmosphere (Metzger 1953). It is indeed 'out there' (remote), but the remoteness has no 'value' (in a numerical sense).

If you articulate the visual field, for instance by presenting a statistically uniform arrangement of polka dots instead of a uniform field (Ganzfeld), most observers become aware of a surface (Koenderink et al. 2009). The surface is apparently at a particular depth, because it is 'thin' and 'located', but it is entirely underdetermined what that depth is. The very notion of '(absolute) location in depth' is alien to the awareness. The surface is just—who knows where exactly?

If *determinacy* is the appropriate test, then it would appear that accommodation fails this test. Compare and contrast the conclusions of two of the leading studies:

Fisher and Ciuffreda (1988): 'The present study demonstrated that information derived from the accommodative system can be used in the judgment of egocentric distance'.

Mon-Williams and Tresilian (2000): 'In conclusion, there appears to be little support for the notion that accommodation provides useful distance information'.

In both studies, subjects indicated the location of a target (placed between 16 cm and 50 cm) by moving their finger underneath or alongside the apparatus. The size of the target was controlled using a Badal lens (in Fisher & Ciuffreda) or slight variations in target size (in Mon-Williams & Tresilian). Mon-Williams and Tresilian confirmed Fisher and Ciuffreda's experimental results but came to a quite different conclusion because, unlike Fisher and Ciuffreda, they chose not to *average* over their results. They found that observers were almost never within 2 cm of the target, and the unsighted errors for the two observers were 39.47 cm (vs. 6.25 cm for full-cue conditions) and 40.49 cm (vs. 7.12 cm for full-cue conditions) even though the full range of possible target distances was 24 cm. So there was no real sense in which subjects were able to use accommodation to determine the distance of the target.

But I don't think *determinacy* is the appropriate test. More intriguing is the possibility (also explored by Mon-Williams & Tresilian) that subjects might be able to use accommodation to judge the ordinal depth of two successive views of the same stimulus at different distances. Mon-Williams and Tresilian found that observers were correct on the majority of trials, which they took to be conclusive evidence that subjects had access to this information. Although Mon-Williams and Tresilian (2000) is generally interpreted as tantamount to a dismissal of accommodation as a depth cue, this is a surprisingly strong result and certainly runs counter to my observations on Optometry.

But is there another possible explanation for this performance? Mon-Williams and Tresilian are concerned with *blur-driven* accommodation: subjects accommodating to a target and then judging its distance. But might the degree of initial defocus blur in the target (either gauged initially or from the watching the target come into focus) provide a visual clue to ordinal depth? This seems more in keeping with the pre-Fisher and Ciuffreda (1988) literature according to which, as Nguyen et al. (2005) recount, subjects could not judge the distance of an object based on accommodation but could use *changes* in accommodation to detect *changes* in distances.

But how might changes in defocus blur enable such a determination? Well, take Fisher and Ciuffreda who test blur-driven accommodation from a *tonic* (or *resting*) state. The tonic accommodation reported (2D) equated to the maximum distance of the target (50 cm). So when estimating the ordinal depth of targets located between 16 cm and 50 cm subjects only needed to gauge the extent of the defocus blur in the stimulus. Indeed, the actual linear relationship that Fisher and Ciuffreda found between the *apparent distance* (y) and *accommodation* (x) in dioptres is suggestive of just such a strategy:

$$y = 0.27x + 2.33$$

For instance, you might ask: Why does accommodation (x) account for only $1/6$ to $1/3$ of the overall value of y, and where does the 2.33, which accounts for $2/3$ to $5/6$ of y, come from? Well, it is tempting to read the 2.33 as the 2D of tonic accommodation; in which case, the apparent distance is tonic accommodation minus some distance ($0.27x$) depending on the degree of blur seen in the stimulus.

But Fisher and Ciuffreda appear to explicitly reject this hypothesis, or at least Ebenholtz's (1981) hypothesis that tonic accommodation acts as a perceptual anchor, because they test tonic accommodation separately using a luminous white surface and it *doesn't* obey the $y = 0.27x + 2.33$ relationship: subjects chose values closer to 29 cm than the 35 cm one would expect. But I would argue that this only confirms the *blur-driven* hypothesis:

First, given the close bunching of the subjects' distance estimates of the luminous surface (29 cm), it begins to look like a strategy to pick a middle value in the face of an indeterminate stimulus. Indeed, Fisher and Ciuffreda appreciate that similar strategies arise in the context of pin-points of light (Gogel and Tietz 1973) and blank luminous forms (Gogel and Da Silva 1987).

Second, since tonic accommodation (2D) now corresponds to two apparent distances (29 cm in the context of the luminous surface and 35 cm in response to the experimental stimulus), Fisher and Ciuffreda conclude: 'Tonic accommodation therefore appears to have no definitive perceptual referent'. But this is a startling conclusion to come to given that the whole purpose of their paper is to suggest that accommodation *is* a depth cue. It amounts to saying: '2D of accommodation has no determinate depth'. Indeed, this should be especially concerning for Fisher and Ciuffreda since 2D is the *only* distance they test using two different experimental paradigms (blur-driven accommodation and tonic accommodation).

Third, in a subsequent paper Fisher and Ciuffreda (1989) altered the tonic accommodation of their subjects by having them focus on a near or far distance for 10 mins (the hysteresis effect) with two interesting observations: First, the estimated distance of tonic accommodation (tested via a luminous surface) did *not* differ significantly between the two conditions even though tonic accommodation itself was set at 3.13D (32 cm) for the near condition and 2.14D (47 cm) for the far condition. Second, the estimated distance of a target placed at 3D (33 cm) reflected the blur-driven change in accommodation; the target was judged to be further in the near condition, and nearer in the far condition, leading Fisher and Ciuffreda to conclude:

> The results suggest that accommodative distance information is linked to the level of monocular blur-driven accommodative innervation necessary for clarity of focus.

But no physiological explanation (innervation) is required; this result could just as easily be explained by the initial blur in the target so long as subjects were aware of the direction they were accommodating (which at these close distances might well be provided by the feeling of tension or relaxation in the ciliary muscles).

Is there any other evidence in favour of accommodation as a depth cue? Well, in the context of *monocular* VR displays, Liu et al. (2010) tested whether subjects were able to determine the location of a virtual target at one of three distances: (a) near (5D = 20 cm), (b) middle (3D = 33 cm), or far (1D = 1 m). On average, subjects judged correctly on four out of five trials (although there was great inter-subject variability). But again, so long as subjects were (a) aware of the direction in which they were accommodating, and (b) started from roughly the same tonic accommodation on each trial, this coarse level of discrimination (near, middle, and far) could be attributed to the stimulus blur.

Furthermore, work in the context of *binocular* VR displays (Watt et al. 2005) provides evidence against accommodation as a distance cue. Watt et al. reasoned that the residual flatness in binocular VR displays might be due to accommodation as a distance cue: when subjects shift their gaze around the display, accommodation might register that all of the visual field is at the very same physical distance. And yet, when Watt et al. asked their subjects to estimate slant whilst keeping their gaze fixed (thereby controlling for accommodation), they found no increase in the slant estimates.

But how might we test accommodation as a depth cue whilst controlling for the blur in the stimulus? Well, a Badal lens magnifies the stimulus at exactly the same rate that perspective shrinks it, enabling us to keep the size of the stimulus fixed as we move it backwards and forwards in space (Fig. 2).

Badal lenses (Fisher and Ciuffreda 1988) and spherical mirrors (Liu et al. 2010) have been used to control for the target size when testing

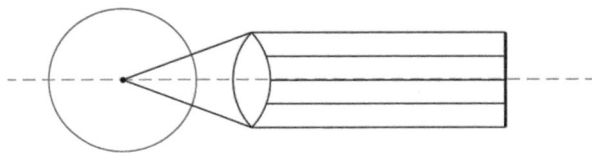

Fig. 2 Schematic of a Badal lens: The focal point of a positive lens (centre) coincides with the nodal point of the eye (left) meaning that the display (right) casts the same sized retinal image whatever its distance.

blur-driven accommodation (a spherical mirror functions in the same way as a Badal lens, but without inducing chromatic aberration). But we could equally ask: (a) What would it be like to view a target moving back and forth in a Badal lens system? Would we see it moving towards us and away from us in depth (indicating that accommodation is a depth cue)? Or would it appear to hover in space at the very same location (indicating that accommodation has no effect upon its perceived depth)? And (b) where exactly would the target be located in space? If accommodation isn't a depth cue, and all the visual system has to go on are the parallel principal rays, the target could be located anywhere from just behind the lens to the horizon.

Having tested this informally, I would tentatively suggest that the Badal lens does appear to break the link between physical distance and perceived depth: As the physical target is moved backwards and forwards in space, the target appears to hover just behind the lens or, to the extent it might appear to move (since the Badal compensation is not perfect, even if it is perfectly aligned), one gets the impression of an object stuck in treacle, resisting any movement, and leading to a powerful disconnect between vision and motion if the target is moved by hand. This illusion has to be formally confirmed but if it is correct then it would provide convincing evidence that previous reports of accommodation as an ordinal depth cue rested upon the initial blur in the stimulus.

For completeness sake we might also consider a Badal lens in reverse, namely a stand magnifier: for instance, an 8D lens on a 12.5 cm stand so that any text placed underneath the stand is automatically located at the focal length of the lens (Fig. 3).

Stand magnifiers are typically used to magnify text for those with poor visual acuity ('low vision'), but the question for us is what this magnification

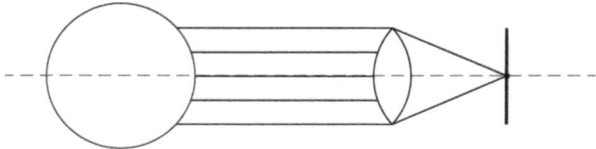

Fig. 3 Schematic of a stand magnifier: A display (right) is placed at the focal length of a positive lens (centre) producing parallel rays. The retinal image is therefore the same size whatever the distance of the eye (left) from the display.

does to the *distance*, rather than the *size*, of the text? And there appear to be three possible answers: the text appears (a) *further away*, (b) *closer*, or (c) *at the same location?*

Now although (a) *further away* may sound paradoxical, it is the answer we would expect if accommodation were a depth cue since an object located at the focal plane of a stand magnifier gives rise to an image at infinity. But do we really have an impression of an image at infinity, or even on the horizon, when we look through a stand magnifier *monocularly?* And by setting vergence at zero, we can ask the same question *binocularly* as well (indeed, one wonders how the vergence-accommodation link responds to a stand magnifier): With (a) vergence set at infinity (i.e. parallel) and (b) accommodation set at infinity (i.e. 0), we *still* don't get an impression of a massive object located on the horizon. This should be no surprise: many stereoscopes rely on uncrossed fusion (i.e. parallel vergence) with lenses setting accommodation at infinity, and yet their whole purpose would be frustrated if these viewing conditions suddenly located objects on the horizon. So *infinity* provides a very useful test case for accommodation and vergence as depth cues.

By contrast, support for the (b) *closer not larger* interpretation comes from Koenderink et al. (2013) who argue that although binoculars are usually stamped with a magnification, in reality their users do not experience a magnification but diminished distance instead. So the choice is between (b) *closer not larger* (Koenderink et al.) or (c) *larger not closer* (magnification). And you might think that Koenderink et al.'s *closer not larger* position has an intuitive appeal: for instance, when we look at the craters of the moon through a telescope they don't appear to be hundreds of thousands of miles away.

But we can test this hypothesis by focusing on a point with both eyes and placing a magnifying lens in front of one eye but not the other (if you are concerned about vergence, focus on a point in the far distance): As you move the magnifying glass from side to side the magnified image appears to shimmer around the unmagnified image but (and this is the important point) they both appear to be located on the very same plane in space. A corollary of this observation is that although new spectacles from an optometrist can make objects in the world seem *larger* (for positive lenses) or *smaller* (for negative lenses) due to the magnification of the lenses (itself a function of the distance of the lens from the eye, which can be eradicated using contact lenses), this should not contribute to these objects *being seen*, as opposed to merely *being judged*, as closer or further away.

1.4   Phenomenal Geometry Without Scale

The discussions from the previous two sections would appear to lead us to the same conclusion, namely that in *monocular* stereopsis we experience phenomenal geometry *without scale*. Indeed, in light of my observations on vergence, I would be tempted to extend this conclusion to *binocular* vision as well. Instead, the scale would appear to be a non-perceptual cognitive inference that we *attribute* to the visual scene. But it is all very well coming to this conclusion in the abstract, how are we make sense of it?

The first thing to notice is that this suggestion isn't nearly as novel as it might at first appear: Although we have primarily focused on *monocular* stereopsis, a similar concern has plagued the *binocular* stereopsis literature for decades in light of the realisation that binocular disparity (at least in Random-Dot Stereograms with no vertical disparity) provides only *relative* depth information. Julesz (1995) toyed with this concern for much of his career:

> ...the question of absolute depth in stereopsis is rather enigmatic.

> Except for echo-locating animals, can primates perceive absolute depth? For instance, stereopsis yields only relative depth.

> ...I am still uncomfortable with the notion that absolute depth can be obtained by the mind...

Nor did Julesz believe that this question could be answered by introspection:

> In the case of another qualia question (Do we sense absolute depth, or only relative depth?), the problem of the sensation of plasticity is as impenetrable as the essence of the sensation of 'color'...

In spite of these concerns Julesz went on to endorse an absolute depth account. But the problem resurfaced in the late-1980s/early-1990s when it was realised that binocular stereopsis failed to preserve the shape of objects as their distance increased. Whilst Johnston (1991) suggested that the problem was an inability to accurately scale binocular disparity using the available depth information, Morgan (1989) began to wonder whether the failure of shape constancy was really a *failure* in the first place? Anticipating the *cognitive turn* (see Chap. 1) Morgan suggested that a central function of stereopsis is precise hand-eye coordination, but since the hand is likely to be in view as we coordinate it Morgan asked whether the primary goal of stereopsis really has to be the extraction of absolute, as opposed to relative, depth?

But how should we begin to make sense of vision *without scale*? I would argue that a good place to start is with Gogel's (1956) 'equidistance tendency': the tendency to see objects as being located at the same distance when all cues as to their relative position have been eradicated. Although Gogel embellished his theory in various ways (see Gogel 1965, 1978, 1990, 1993), his fundamental observation was an elaboration of Emmert's law (1881) which suggests that the perceived size of an after-image is in direct proportion to the distance of the real-world object upon which the after-image is being projected. Gogel would explain Emmert's law in the following terms: after-images are (i) completely impoverished so far as depth cues are concerned, but (ii) two objects placed side by side will appear to be at the same distance (the 'equidistance tendency'), so (iii) since the real-world object *does* have a determinate distance, the distance of the real-world object *captures* the distance of the after image.

But notice that this whole account is articulated in terms of *physical distances*. Now imagine that vision has no scale. Instead, the plane of fixation *always* corresponds to a *single fixed phenomenal depth*, and so whatever is being fixated upon, whether it is near or far, is seen as being at that *single fixed phenomenal depth*. In which case Gogel's account would be turned upside down: It would be the depth of the after-image that remained fixed, and it would be the depth of the real world that alters as we shift our fixation from one real-world object to the next. This is, after all, still entirely consistent with Emmert's and Gogel's observations.

It is also consistent with our experience stereoscopes: our enjoyment of scenes of vastly different scales is not impeded by the fact that accommodation and vergence always remain fixed in a stereoscope and typically specify a distance on the horizon. It is true that vertical disparities might be present, but for many stereograms they are not (see Howard & Rogers, 2012) and in any case it is unclear the extent to which vertical disparities contribute to our perception of distance: the only evidence of vertical disparities scaling the distance of a fronto-parallel surface appears to be Appendix A of Rogers & Bradshaw (1995) and this effect is (a) contingent upon the stimulus taking up at least 20° of the visual field, and (b) goes hand in hand with the fronto-parallel surface appearing increasingly curved at near distances. But given the fall-off of binocular disparity with distance we might reasonably expect subjects to judge curved surfaces with stereoscopic depth as being closer than flat surfaces irrespective of whether distance is conveyed perceptually or not.

Finally, isn't my *single fixed phenomenal depth* simply Gogel's (1969) *specific distance tendency*: the tendency to see objects as being located at roughly 2 m in the absence of any distance cues to the contrary? No, for two reasons: First, I regard the *specific distance tendency* as merely a decision strategy adopted in a context of uncertainty. As Howard and Rogers (2012) observe:

> Contraction of judgments towards the mean is not unique to distance judgments. It occurs in any task when there is uncertainty (Mon-Williams et al. 2000). One would expect that judgments of distance in the absence of depth information would tend to the most probable distance in natural scenes. Yang and Purves (2003) found that, in natural scenes, the probability distribution of physical distances of objects has a maximum at about 3 m.

Second, trying to judge *distance* from a *fixed phenomenal depth* would be a category mistake: *Distance* and *phenomenal depth* are completely divorced under my account. To go back to the example of a stereoscope with fixed accommodation and vergence: whether we are viewing a small object up close or looking out over the horizon, both scenes ought to be experienced as being at the very same phenomenal depth even though we may cognitively attribute vastly different scales to them.

2   PHENOMENAL GEOMETRY

The second question this chapter seeks to address is what determines the perceived geometry of the scene when binocular disparity is absent? Specifically, can a *purely optical* account of depth perception explain the perceived variation in depth across a monocular scene?

The obvious candidate for an *optical* cue to depth variation is defocus blur. Following Pentland (1987) in computer vision, the last 20 years has seen an increased consideration of defocus blur as a potential depth cue (see Marshall et al. 1996; Mather 1996, 1997; O'Shea et al. 1997; Mather and Smith 2000, 2002; Palmer and Brooks 2008). Indeed in the last 10 years, largely thanks to the work of the Banks Lab, defocus blur has been promoted from being a 'relatively coarse, qualitative depth cue' (Mather and Smith 2002) to a central concern modern VR displays (see Watt et al. 2005; Held et al. 2012a; Banks et al. 2016 for an overview).

But if we are to understand the contribution that defocus blur makes to depth perception, we first have to consider the two levels at which it might operate: (a) visual experience (*perceptible blur* in the visual field) or (b) optics (*optical defocus* in the retinal image). Whilst the contemporary literature tends to focus on *perceptible* blur, I would argue that the significance of defocus blur might actually be attributable to *imperceptible* defocus in the retinal image.

2.1   Perceptible Blur in the Visual Field

We should therefore distinguish between (a) *perceptible blur*, as a quality of our subjective visual experience; specifically, the fact that objects can be seen as *more or less blurry*, and (b) *optical defocus*, as an objective optical property of the retinal image itself. Indeed, *depth of field* represents the degree of optical defocus that can exist *without* perceptible blur penetrating our subjective visual experience. Depth of field is typically around ± 0.25D to ± 0.5D (see Wang and Ciuffreda 2006) and, if estimates of microfluctuations also being about ± 0.25D to ± 0.5D are correct (see Charman and Heron 2015), may well be the visual system's way of ensuring a sharp visual experience in spite of its constantly fluctuating optics.

But what is interesting, indeed one might say perplexing, about the defocus blur literature is the continual emphasis on *perceptible blur* rather than *optical defocus*, that is on treating defocus blur as just another pictorial cue alongside, say, perspective and shading. This approach to defocus blur appears to originate with a division of labour between Psychophysics and Cognitive Psychology: In Psychophysics, Watt and Morgan (1983) had shown that subjects were quite good at detecting the presence or absence of blur, and so the question left open for Cognitive Psychology was whether subjects attached *depth meaning* to the blur they detected? But with this commitment to *perceptible blur* comes three significant concerns.

a. Infrequency: Perceptible blur is largely absent from much of our visual experience. As Green et al. (1980) observe, according to Campbell's (1957) estimate of the depth of field ($\pm$ 0.43D), all objects from 2.3 m to infinity ought to be in focus for the unaccomodated observer. Indeed, it was for this reason that Gibson (1950) rejected defocus blur as a basis for his optical account of depth perception, calculating (with the aid of Mohler) that all of the visual field apart from that directly under the nose tends to be in uniform focus. And in the contemporary literature, Vishwanath (2012a, b) appeals to Wang et al.'s (2006) work on subjective 'equiblur' zones to make a similar point in response to Held et al. (2012a, b).

b. Unreliable: The psychophysical literature on the discrimination of defocus blur suggests that it can only function as the coarsest of depth cues: First, once we try to go beyond merely *detecting* the *presence* blur (depth of field) to *discriminating* the *degree* of blur, it turns out that subjective thresholds are actually quite high—a significant shift in the physical location of an object is required before a change in visual blur is detected (Mather and Smith 2002). Second, subjects appear only to be able to detect four discrete levels of blur in a single scene (Taylor and Bex 2015). Third, some stimuli require five times more optical defocus than others in order to achieve the very same degree of perceptible blur (Sebastian et al. 2015).

c. Ineffectual: The empirical literature from the mid-1990s to the early-2000s found that subjects were surprisingly poor at utilising the information available from blur when making ordinal depth judgements. Consider border-blur (Marshall et al. 1996; Mather and Smith 2002): If the border between a blurred surface and a sharp surface is *sharp* this indicates that the occluding edge belongs to the sharp surface and therefore the blurred surface is *behind* the focal plane, whilst if the border is *blurred* this indicates that the occluding edge belongs to the blurred surface and therefore the blurred surface is *in front* of the focal plane. Yet subjects are surprisingly poorly at this task, something that Zannoli et al. (2016) have recently confirmed.

But for Sprague et al. (2016) studies of blur in the laboratory miss the point that if blur is a pictorial cue to depth like perspective or shading then (just like perspective and shading) it has to be interpreted in light of natural scene statistics, i.e. asking *when* visual blur is most likely to be encountered in the visual field and *why?* So whilst subjects may be poor at extracting ordinal depth from defocus blur in laboratory conditions,

Sprague et al. argue that in their everyday lives subjects appear to rely on two rules of thumb to disambiguate defocus blur:

1. A vertical blur gradient is interpreted as surface slanted away from you:/, rather than a surface slanted towards you:\.

2. Sharp objects are seen as near, and blurred objects as far.

And both of these rules of thumb are consistent with the scene statistics observed by Sprague et al. in four everyday tasks (walking outside, walking inside, ordering a coffee, and making a sandwich), leading Sprague et al. to conclude that subjects do in fact rely upon natural scene statistics in order to disambiguate defocus blur.

But natural scene statistics also confirms a primary concern for their account, namely the *infrequency* of perceptible blur. Sprague et al. observe that detectable blur is surprisingly uncommon: 'In the weighted combination, blur values exceed threshold less than 4% of the time'. Indeed, subjects appear to adopt a fixation strategy that effectively eradicates perceptible blur from their visual experience with Sprague et al. documenting how detectable blur would be much more common for random fixations than it is was for the actual fixations recorded from subjects as they engaged in everyday tasks. Sprague et al. recognise that this poses a significant challenge for their account: 'If blur were never detectable in natural vision, it would not be a useful depth cue'. Nonetheless, Sprague et al. attempt to refashion this shortcoming into a strength by arguing that although visual blur is rarely encountered in the visual field, it is now especially informative because it is so rare and so 'means there is something unusual about the current viewing parameters and the scene one is viewing...'.

And this would be a suitable conclusion to come to were it not for two concerns: First, the Banks Lab and I have very different aspirations for defocus blur—whilst they are simply looking for another pictorial cue to depth, I am looking for the single optical cue that determines the perceived geometry of monocular stereopsis. Consequently, the fact that monocular stereopsis often occurs in the absence of perceptible blur indicates (so far as my account is concerned) that we still haven't gotten to the heart of the matter. Second, whilst this explanation may account for the performance of subjects in response to *simulated* defocus blur, it still doesn't account for the Banks Lab's own results in the context of *natural* (or *unsimulated*) defocus blur: here subjects perform well in judging ordinal depth from occlusions when the blur is produced by their own visual system in contrast to their performance in response to *simulated* blur (see Zannoli et al. 2016), and natural scene statistics cannot explain this difference in performance.

2.2   Optical Defocus in the Retinal Image

The solution, I suggest, is to shift our attention from perceptible blur in the visual field to subthreshold optical defocus in the retinal image. For instance, Mather and Smith (2002) contrast visual blur as a depth cue with accommodation:

> The poor discriminability of blur indicted by these and other experiments contrasts sharply with the high sensitivity of the accommodative system. Kotulak and Schor (1986) found that an accommodative response can be elicited by a blur stimulus that is below the threshold for blur perception. They concluded that the accommodative system uses a mechanism that does not rely on perceptible levels of blur.

There's no reason in theory why depth from defocus blur couldn't rely on a similar mechanism. But the response that Mather and Smith would give, and indeed the point that they are making in this very passage, is that subjects are unable to respond to perceptible blur so what makes us think that they can take advantage of sub-threshold defocus in the retinal image?

But the significance of the shift from treating defocus blur as a pictorial cue to treating it as an optical cue is that it is now imperative that we are able to *realistically* simulate optical defocus in the retinal image. Painters have long been able to evoke an impression of depth with a rough approximation of visual blur (O'Shea et al. 1997 give Caravaggio's *Boy with a Basket of Fruit*, c. 1593 and Vermeer's *The Lacemaker*, c. 1669–1670 as early examples), but this will no longer suffice: the question is how we can *realistically* simulate optical defocus in the retinal image, and whether such simulations are able to bring subjects back to their real-world performance?

a. Gaussian Blur: Gaussian blur is the standard technique for simulating blur in the literature, and Gaussian blur gradients have been shown to be effective pictorial cues to distance (see Fig. 1) and shape/slant (see Sprague et al. 2016), although blur gradients are liable to be double-counted (for instance, Vishwanath and Blaser 2010 found that when a blur gradient is applied to a surface, it 'appears substantially closer and appears more slanted.').

But Langer and Siciliano (2015) found that when Gaussian blur is applied to *individual objects* in the visual field, rather than as a blur *gradient*, it had little effect on ordinal depth judgements: for their higher blur values performance was at chance, and for their lower blur values performance was only marginally above chance. This was a surprising result because Langer and Siciliano were replicating an earlier study (Held et al. 2012a) where subjects had performed well in response to *natural* (i.e. *unsimulated*) blur (blur produced by the subject's own visual system) (cf. Vishwanath 2012a; Held et al. 2012b; Vishwanath 2012b); the only difference was that the blur was *simulated* in Langer and Siciliano. Langer and Siciliano wondered what was missing from their simulated blur and concluded that optical aberrations, specifically chromatic aberration, were the most likely candidate:

> There is also evidence that chromatic aberration can be used for depth discrimination (Nguyen, Howard, & Allison, 2005). Indeed a recent computational model has shown how chromatic aberration, as well as detailed spatial properties of the eye's point spread function such as astigmatism could allow the visual system to obtain quite accurate estimates of depth (Burge & Geisler, 2011).

> Can these 'higher order' blur cues be the reason why our experimental findings differed from those of Held et al?

b. Blur Circle: But it has increasingly been suggested that Gaussian blur might itself be the cause for concern. For instance, Sebastian et al. (2015) argue that Gaussian blur is unlikely to reflect defocus blur produced by the human visual system, and so only rely on natural (or unsimulated) blur for their psychophysical experiments. By contrast, Zannoli et al. (2016) attempt to simulate defocus blur, but look for a non-Gaussian alternative. Zannoli et al. observe that according to geometric optics the blur circle for a point is a circular disc of equal luminance not a Gaussian, and they follow Potmesil and Chakravarty (1982) and Cook et al. (1984) in this regard (as Barsky et al. 2003 observe of Potmesil & Chakravarty: 'Each pixel is smeared into a disc of a particular radius known as the *circle of confusion* to establish blur.').

The other innovation that Zannoli et al. embrace is Cook et al.'s (1984) distributed ray tracing: since Zannoli et al. are concerned with testing blur at an occlusion edge, they use ray tracing to model how the blur circle from *myopia* (i.e. blur from the surface that is being occluded) is truncated into a semicircle by the occluding edge and produces an astigmatic depth of field.

And yet in spite of these improvements in their stimuli Zannoli et al. found that the performance of their subjects in their *simulated* blur condition (with a single-plane display) was only marginally above chance, in contrast to the subjects' performance in the *natural* blur condition (where blur was the product of the subject's own visual system by setting one plane in front of the other in a multiplane display). Zannoli et al. wonder what was missing from their single-plane display that was present in their multiplane display? And, as with Langer and Siciliano (2015), chromatic aberration emerges as the prime candidate:

> We asked what the critical differences were in the single- and multiplane cases. We found that chromatic aberration provides useful information but accommodative microfluctuations do not.

c. Optical Defocus: But before considering whether chromatic aberration really is the essential missing ingredient, we could further improve our single-plane stimulus by trying an intermediate position between (a) relying on computer simulation (which *doesn't* work, see Langer and Siciliano 2015; Zannoli et al. 2016) and (b) relying on the optics of the human eye (the multiplane condition in Zannoli et al. 2016, which *does* work), namely, by (c) relying on optical defocus that is not produced by the human eye. For instance, Held et al. (2010) speculated that if we wanted to replicate the pattern of defocus blur on the retina, all we would need to do is photograph the scene using an aperture the same size as the viewer's pupil: 'If a viewer looks at the resulting photograph from the center of projection, the pattern of blur on the retina would be identical to the pattern created by viewing the scene itself'.

Maiello et al. (2015) recently used a Lytro plenoptic camera to test something close to this hypothesis. The camera captures 12 images of the same scene simultaneously, each with a different focal length, so Maiello et al. could ensure (a) that the only difference between the images was defocus blur, and (b) that the defocus blur in the stimulus was *optically produced* rather than *simulated*. And yet, Maiello et al. found that the presence of this blur only impeded their subjects' ability to judge the ordinal depth of image patches of natural scenes. Again, as with Langer and Siciliano (2015) and Zannoli et al. (2016), Maiello et al. wonder why their blur stimulus was ineffective. And again, as with Langer and Siciliano (2015) and Zannoli et al. (2016), the absence of chromatic aberration emerges as the most likely candidate:

> The main difference between our study and the previous is that Held et al. employed a volumetric display to present observers with blur due to the optics of their own eyes. Thus, in the Held study, observers might have been able to employ other optical cues, such as chromatic aberration, that have been shown to be possible cues to depth sign. [fn. Nguyen et al. 2005]

So all three of the recent studies on the ineffectiveness of simulated defocus blur as a depth cue regard chromatic aberration as an essential missing ingredient. Indeed, all three cite Nguyen et al. (2005) as evidence in favour of this proposition. Taken at face value, the results in Nguyen et al. do seem suggestive: a complete drop-off in the ability to judge the ordinal depth of two vertical planes staggered in space when the planes were viewed under monochromatic light. But there are good reasons to be sceptical of this conclusion:

First, Nguyen et al.'s ordinal depth study was inspired by similar work in accommodation where monochromatic light was found to disrupt the direction of the accommodative response (which implicitly relies upon an ordinal depth judgement between the focal plane and the intended target). But it is well documented within the accommodation literature that the drop-off in performance is far from absolute, see Wang et al. (2011). So it should concern us that Nguyen et al. appear to find a *complete* drop-off in performance even though this is explicitly *not* what would be predicted by the accommodation literature.

Second, is a *complete* drop-off in performance under monochromatic light even what Nguyen et al. really find? Nguyen et al. tested five subjects: (a) Only three of the subjects were able to perform above chance under *white light* with *no time constraints*, so for two subjects there was no drop-off. (b) Going to the other extreme, only one of the subjects (VN) was able to perform above chance under *white light* with *a brief exposure* (0.21 s). But then VN performed equally well with a *brief exposure* under *monochromatic light*, so again no drop-off here. (c) So the only evidence for a drop-off is the fact that three of the five subjects (including VN) were at chance under *monochromatic light* with *no time constraints* even though they had been above chance under *white light* with *no time constraints*. Admittedly it is strange that VN should lose their ability to make ordinal depth judgements under monochromatic light with *no time constraints* but not under monochromatic light with *a brief exposure*, but VN's poor performance with *no time constraints* cannot be used to eradicate their performance with *a brief exposure*.

Third, even the drop-off in performance under *monochromatic light* with *no time constraints* disappears once subjects are permitted to vary their focus back and forth between the two edges. But how are we to make sense of this improved performance? Nguyen et al. suggest that subjects must either be relying on dynamic changes in image blur or afferent signals in the ciliary muscles. But we have already explored why we might be sceptical of afferent signals in the ciliary muscles, leaving us with dynamic changes in image blur as the only possible explanation. But why should this help subjects determine the *order*, as opposed to merely the *extent*, of the depth separation between the two edges?

In conclusion, I'm not claiming that chromatic aberration makes *no* contribution to depth perception, only that it is not *in and of itself* the essential ingredient that differentiates natural from simulated blur. But instead of thinking of chromatic aberration as a distinct source of depth information, perhaps we should think of it merely as *accentuating* or *amplifying* other sources of depth information? All the optics of chromatic aberration mean when conjoined with the rod and cone physiology of the retina is that we have *four* retinal images rather than one, effectively at four different degrees of optical focus/defocus. So perhaps chromatic aberration matters because it gives us *four* snapshots of another source of information: one snapshot by itself might be enough to (a) determine the direction of accommodation or (b) determine ordinal depth, but four snapshots together provide an even more accurate impression, thereby explaining why chromatic aberration appears to have a *significant impact* on accommodation and ordinal depth without functioning as a *necessary condition*.

This interpretation would be entirely consistent with the findings in Zannoli et al. (2016), in contrast to their hypothesis that 'the nearly correct handling of chromatic aberration in the multiplane display provided the additional information required for depth-order judgments'. In order to confirm that chromatic aberration is the missing source of information, Zannoli et al. attempted to present their multiplane stimuli without chromatic aberration. But unlike the Low-Pressure Sodium (SOX) lamps of Nguyen et al. (2005), the multiplane stimuli in Zannoli et al. were presented on CRT (TV) display meaning that chromatic aberration could only be *reduced* (by showing the stimuli on the green channel of the CRT) rather than *virtually eradicated*. Consequently, Zannoli et al. found what was, by their own admission, only a small reduction in performance which was some distance from confirming chromatic aberration as the essential missing source of information.

But if chromatic abberation isn't entirely to blame, then what other sources of information might contribute to the difference in performance between Zannoli et al.'s *natural* and *simulated* blur conditions?

a. Microfluctuations: Small fluctuations in the optical power of the visual system have two components: a low-frequency component (2 s per cycle) and a high-frequency component (0.4–0.8 s per cycle). The low-frequency component is the more plausible visual cue not only because it is an order of magnitude larger but also because, as Charman and Heron (2015) observe, 'most authors agree that the LFC are under neurological control, since they vary systematically with task and observing conditions'. Although Charman and Heron raise the concern that microfluctuations are simply too slow to inform accommodation, (a) there is no reason why a full 2 s cycle should be required before directional information could be extracted from a dynamic system, and (b) in any case microfluctuations might be constantly working in the background to contribute to our perception of depth, and it might be this depth perception rather than the microfluctuations themselves that directly informs accommodation.

But the greater concern is that when Zannoli et al. applied a cycloplegic drug to paralyse these microfluctuations they found that this had no effect on subjects' abilities to determine ordinal depth. Zannoli et al.

recognise that their cycloplegic drug may only control for low-frequency microfluctuations since the high-frequency component is an artefact of the subject's pulse. Furthermore, Zannoli et al. admit that the cycloplegic drug may only greatly reduce the high-frequency component rather than completely eradicate it. This leaves open the slender possibility that small variations in defocus induced by the high-frequency component and/or a residual low-frequency component might be enough to indicate that the *natural* blur is not static in the way that the *simulated* blur is.

b. Monochromatic Aberrations: Zannoli et al. recognise that *natural* blur is subject to a whole host of other aberrations (astigmatism, coma, trefoil, and spherical aberration), but suggest that these aberrations cannot explain why *simulated* blur is regarded as artificial since *simulated* blur is also subject to these aberrations:

> The image of a real scene is affected by defocus blur as well as by diffraction and higher order aberrations. As the eye becomes more and more defocused, the defocus component becomes the dominant effect, while the other effects remain roughly constant (Wilson et al. 2002). To a first approximation, the latter effects are independent of defocus, so the total blur is a combination of the two. Then, assuming that participants accurately focus on the stimulus, one should add only blurring due to simulated defocus, because the other effects will be inserted by the viewer's eye.

But there are two concerns with this passage:

First, when Wilson et al. (2002) showed subjects myopic and hyperopic blur produced in the presence of monochromatic aberrations the subjects were able to correctly identify whether the blur was myopic or hyperopic in 83% of the subsequent trials after just 2 mins of training. This information is lost in Zannoli et al.'s stimulus since they use an identical stimulus to simulate myopic and hyperopic blur: a circular disc of uniform intensity (although they take into account how the occlusion border truncates the blur circle in myopia). There is a lively debate as to the extent to which these monochromatic aberrations contribute to ordinal depth: as we have already seen the accommodation literature would appear to cast doubt on the majority of these monochromatic aberrations, but see Lopez-Gil et al. (2016) on the possibility that retinal blood vessels differentiate myopic and hyperopic blur.

Second, even though *simulated* blur is subject to the same monochromatic aberrations as *natural* blur, it is not subject to them in the same way. Take the case of myopic *natural* blur with positive spherical aberration: All of the rays from a point source are equally over-focused by myopia, but then the addition of spherical aberration means that the rays that pass through the outer edge of the lens, and therefore form the outer edge of the blur circle, are more over-focused than the rays that pass through the centre of the lens. But imagine the blur circle is now *simulated* with a 2D circle: Every point in the 2D image of the circle now becomes its *own* point source, with its *own* point spread function from spherical aberration. So, spherical aberration no longer accentuates the *natural* defocus of a single point in certain directions; instead, it degrades all of the individual points of a *different* stimulus: the circle specified by *simulated* blur. This radically alters the resulting retinal image, and if the visual system uses this to deduce that *simulated* blur is made up of independent point sources, rather than the over- or under-focused rays from a single point source, this could explain why *simulated* blur is interpreted as artificial.

c. Resolution: Zannoli et al. are concerned with *above*-threshold blur. To the extent that the visual system may be responsive to *sub*-threshold blur, concerns about image resolution become greater, specifically with regards to (a) the resolution of the display itself, and also (b) the sampling method used to render the blur (Zannoli et al. rely on simulating defocus for just 100–200 points in the image).

d. Asymmetrically Tilted and Decentred Optics: Similarly, to the extent that the visual system may respond to sub-threshold blur, the greater concerns about the asymmetrically tilted and decentred optics of the eye become (see Polans et al. 2015). So, too, might the variation in pupil shape with eccentricity (a circular pupil becomes increasingly oval with eccentricity from the optical axis, see Thibos and Liu 2016) especially since the visual axis is already offset from the optical axis, although given Zannoli et al.'s field of view is $10°$ this may be of negligible concern.

e. False Positive? Finally, we have to be open to the possibility that Zannoli et al. provide us with a false positive in their *natural* blur condition. Although the blur in their *natural* condition is not simulated, the occlusion border that subjects have to judge is: see Narain et al. (2015).

2.3   *Optical Distortions*

Zannoli et al. (2016) explore whether blur at an occlusion border can be used to judge which of two surfaces is in front of the other. But do occlusion borders do more than simply indicate the *order* between two surfaces? Pinch your forefingers and thumbs together to form a 1 cm aperture and view a page of text on a computer screen: as you move the aperture around, the words on the page appear subtly distorted. Looking at a grid through the aperture appears to suggest that the aperture induces a subtle pincushion distortion. You can accentuate the effect by narrowing the aperture (i.e. pinching your thumbs and forefingers closer together) or increasing the distance of the aperture from your eye. The effect also persists if we use an iris aperture or simply a hole punched in a piece of card.

One commentator has suggested to me that this might be the visual system's interpretation of the 'vignetting' (or luminance gradient) produced by the blurring of the aperture's edge. Another possibility is that the partial visibility of the computer screen through the blurred aperture inclines the visual system to try and stitch the two surfaces together, see Akeley et al. (2004). But vignetting and pincushion distortions typically go hand in hand when there is an *optical* explanation for the distortion in question (see, e.g. x-ray machines with an image intensifier). And one possible *optical* explanation for this distortion is that the points on the display that are just next to the aperture are blocked out for the majority of the eye and are only focused by the extremity of the intraocular lens, explaining why (a) the image is reduced in luminance (since it is missing the majority of the light rays), but also (b) why the image is magnified (since the intraocular lens is more powerful at its edges). It might also explain the miniaturisation that subjects in Vishwanath and Blaser (2010) and Vishwanath and Hibbard (2013) report from aperture viewing (although Vishwanath 2014 attributes this miniaturisation to his depth-scaling hypothesis): For the centre of the scene, the image is only focused by the centre of the intraocular lens.

Given that 1 cm apertures placed 1.5–2 cm in front of the eye are often used in experiments to block out the borders of displays, we have to be aware of the potentially distortive effects that such apertures may have on stimuli. For instance, Volcic et al. (2014) employ *two* apertures when testing the monocular stereoscopic impression from 2D images: (a) a 1 cm aperture 1.5–2 cm from the eye and (b) an additional 12.5 cm aperture 40 cm from the eye 'to block the view of the monitor frame'. Strictly speaking, only (b) should have been required, but would the effect have persisted in quite the same way had (a) been removed?

This discussion of distortions raises a further question: There are optical instruments, such as the Zograscope, whose very purpose is to introduce uneven magnification in a 2D image as a way of inducing a 3D impression of depth (Fig. 4). Koenderink et al. (2013) reject an optical distortion account of the Zograscope as 'evidently nonsensical', with Koenderink (2015) only willing to revise this assessment if you introduce head movements of sufficient magnitude to induce motion parallax. But the Zograscope does not require such head movements, leaving Koenderink (2015) referring to the Zograscope as 'this miraculous 18th-century "optical" instrument that works despite the fact that it lacks a scientific basis'.

Christopher Tyler has suggested to me that the Zograscope might primarily be a binocular device that relies upon the differential distortions projected to the two eyes to induce a curved disparity field. But monocular viewing of the Zograscope does raise one of the central questions of Chap. 2, namely, whether perspective acts as a depth cue? Specifically, is a 2D pincushion distortion *seen* as a 3D concave surface? (The Zograscope also induces differential defocus blur across the 2D image, but in the opposite direction to magnification: Hyperopia specifies miniaturised objects as being in front of the focal plane, whilst myopia specifies magnified objects as being behind the focal plane.)

But if our experience of the Zograscope is explained by the fact that a 2D pincushion distortion is *seen* as a 3D concave surface it would make the Zograscope redundant: exactly the same effect could simply be achieved by introducing these distortions into the 2D image itself. And this is, in effect, what Rogers and Brecher (2007) argue Helmholtz achieved with his distorted chequerboard which looks undistorted when brought close to the eye (at the distance marked 'A' in the diagram) (Fig. 5).

Rogers and Brecher argue that the distorted 2D pattern is interpreted as an undistorted pattern on a concave 3D surface, so we experience a monocular stereoscopic impression of a concave 3D surface. But I would be cautious of this explanation for the simple reason that the chequerboard also looks relatively undistorted when we view it from

Fig. 4 Zograscope for viewing 2D images: A 45° mirror directs light from an etching laid flat on a table through the convex lens and towards the viewer's eyes. Courtesy of Georgian Print Rooms. © Georgian Print Rooms. For more information please, see http://www.georgianprints.co.uk/typesofprints/Zograscope/zograscope.html

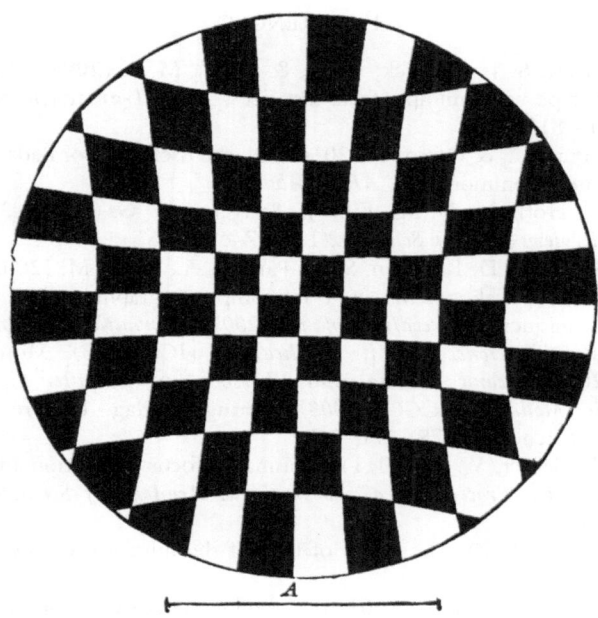

Fig. 5 Helmholtz chequerboard (Helmholtz 1866). Scan courtesy of James McArdle: https://drjamesmcardle.com/2013/04/07/from-the-corner-of-your-eye/

the side, and place the regular squares in the periphery and the magnified squares in our central vision. This would appear to suggest that Helmholtz's chequerboard might simply reflect the fact that at the close viewing distance specified by Helmholtz the extremities of the chequerboard are simply further away from the eye than its centre, consequently as the object shrinks with distance some compensatory mechanism has to be introduced. If this is the correct interpretation then Helmholtz's chequerboard is really no different from the works of Georges Rousse and Felice Varini: pieces of *anamorphic art* that when viewed from the right position produce an impression of a regular pattern in spite of the fact that the surfaces they are painted on are not at the same distance. The test of this hypothesis is whether an anamorphic chequerboard that was constructed for side-on viewing (like William Scrots' portrait of King Edward VI, c. 1546) would look undistorted?

REFERENCES

Akeley, K., Watt, S. J., Girshick, A. R., & Banks, M. S. (2004). A stereo display prototype with multiple focal distances. *ACM Transactions on Graphics, 23*(3), 804–813.

Aldaba, M., Pujol, J., & Otero, C. (2016). On the usefulness of Badal optometer to stimulate accommodation. *ARVO Abstract.*

Banks, M. S., Hoffman, D. M., Kim, J., & Wetzstein, G. (2016). 3D displays. *Annual Reviews of Vision Science, 2*(1), 397–435.

Barsky, B. A., Horn, D. R., Klein, S. A., Pang, J. A., & Yu, M. (2003). Camera models and optical systems used in computer graphics: Part II, image-based techniques. *Proceedings of the 2003 International Conference on Computational Science and its Applications (ICCSA'03)*, Montreal, May 18–21 2003. *Second International Workshop on Computer Graphics and Geometric Modeling (CGGM'2003)*, Springer-Verlag Lecture Notes in Computer Science (LNCS), 256–265.

Burge, J., & Geisler, W. S. (2011). Optimal defocus estimation in individual natural images. *Proceedings of the National Academy of Sciences, 108*(40), 16849–16854.

Campbell, F. W. (1957). The depth of field of the human eye. *Optica Acta, 4,* 157–164.

Charman, W. N., & Heron, G. (2015). Microfluctuations in accommodation: An update on their characteristics and possible role. *Ophthalmic and Physiological Optics, 35*(5), 476–499.

Chen, L., Kruger, P. B., Hofer, H., Singer, B., & Williams, D. R. (2006). Accommodation with higher-order monochromatic aberrations corrected with adaptive optics. *Journal of the Optical Society of America A, 23*(1), 1–8.

Chin, S. S., Hampson, K. M., & Mallen, E. A. (2009a). Role of ocular aberrations in dynamic accommodation control. *Clinical & Experimental Optometry, 92*(3), 227–237.

Chin, S. S., Hampson, K. M., & Mallen, E. A. (2009b). Effect of correction of ocular aberration dynamics on the accommodation response to a sinusoidally moving stimulus. *Optics Letters, 34*(21), 3274–3276.

Cook, R. L., Porter, T., & Carpenter, L. (1984). Distributed ray tracing. SIGGRAPH '84 Proceedings of the 11th Annual Conference on Computer Graphics and Interactive Techniques (137–145).

Cope, J. A. (1993). The physics of the compact disc. *Physics Education, 28,* 15–21.

Ebenholtz, S. M. (1981). Hysteresis effects in the vergence control system: Perceptual implications. In D. F. Fisher, R. A. Mority, & J. W. Senders (Eds.), *Eye movements: Visual perception and cognition.* Hillsdale, NJ: Lawrence Erlbaum.

Emmert, E. (1881). Grössenverhältnisse der Nachbilder. *Klinische Monatsblätter für Augenheilkunde, 19,* 443–450.

Fernández, E. J., & Artal, P. (2005). Study on the effects of monochromatic aberrations in the accommodation response by using adaptive optics. *Journal of the Optical Society of America A, 22*(9), 1732–1738.

Fisher, S. K., & Ciuffreda, K. J. (1988). Accommodation and apparent distance. *Perception, 17*(5), 609–621.

Fisher, S. K., & Ciuffreda, K. J. (1989). The effect of accommodative hysteresis on apparent distance. *Ophthalmic and Physiological Optics, 9,* 184–190.

Frisby, J. P. (2009). Optic arrays and retinal images. *Perception, 38,* 1–4.

Gambra, E., Sawides, L., Dorronsoro, C., & Marcos, S. (2009). Accommodative lag and fluctuations when optical aberrations are manipulated. *Journal of Vision, 9*(6), 4.

Gibson, J. J. (1947). *Motion picture testing and research.* Research Reports, Report No. 7, Army Air Forces Aviation Psychology Program.

Gibson, J. J. (1950). *The perception of the visual world.* Boston: Houghton Mifflin.

Gillam, B. (2007). Stereopsis and motion parallax. *Perception, 36,* 953–954.

Gillam, B. (2009). On frisby 'Three cheers for retinal images'. *Perception, 38,* 162–163.

Gogel, W. C. (1956). The tendency to see objects as equidistant and its inverse relation to lateral separation. *Psychological Monographs: General and Applied, 70*(4), 1–17.

Gogel, W. C. (1965). Size cues and the adjacency principle. *Journal of Experimental Psychology, 70*(3), 289.

Gogel, W. C. (1969). The sensing of retinal size. *Vision Research, 9,* 3–24.

Gogel, W. C. (1978). Size, distance and depth perception. In E. C. Carterette & M. P. Friedman (Eds.), *Handbook of perception, vol. 9: Perceptual processing.* New York: Academic Press.

Gogel, W. C. (1990). A theory of phenomenal geometry and its applications. *Perception and Psychophysics, 48,* 105–123.

Gogel, W. C. (1993). The analysis of perceived space. In S. C. Masin (Ed.), *Advances in psychology, vol. 99: Foundations of perceptual theory* (pp. 113–182). Amsterdam: North-Holland.

Gogel, W. C., & Da Silva, J. A. (1987). Familiar size and the theory of off-sized perceptions. *Perception and Psychophysics, 41*(4), 318–328.

Gogel, W. C., & Tietz, J. D. (1973). Absolute motion parallax and the specific distance tendency. *Perception and Psychophysics, 13,* 284–292.

Green, D. G., Powers, M. K., & Banks, M. S. (1980). Depth of focus, eye size, and visual acuity. *Vision Research, 20,* 827–835.

Gregory, R. L. (2009). Prior beliefs. *Perception, 38,* 163.

Gwiazda, J., Thorn, F., Bauer, J., & Held, R. (1993). Myopic children show insufficient accommodative response to blur. *Investigative Ophthalmology & Visual Science, 34,* 690–694.

Hampson, K. M., Chin, S. S., & Mallen, E. A. (2010). Effect of temporal location of correction of monochromatic aberrations on the dynamic accommodation response. *Biomedical Optics Express, 1*(3), 879–894.

Held, R. T., Cooper, E. A., O'Brien, J., & Banks, M. S. (2010). Using blur to affect perceived distance and size. *ACM Transactions on Graphics, 29*(2), 19.

Held, R. T., Cooper, E. A., & Banks, M. S. (2012a). Blur and disparity are complementary cues to depth. *Current Biology*, March 6.

Held, R. T., Cooper, E. A., & Banks, M. S. (2012b). Response to Vishwanath. Originally published alongside Vishwanath (2012a) online, currently unavailable.

Holmes, G. (1918). Disturbances of vision by cerebral lesions. *British Journal of Opthalmology, 2,* 449–468 and 506–518.

Holmes, G., & Horrax, G. (1919). Disturbances of spatial orientation and visual attention with loss of stereoscopic vision. *Archives of Neurology and Psychiatry, 1,* 385–407.

Howard, I. P., & Rogers, B. J. (2012). *Perceiving in depth*. Oxford: Oxford University Press.

Johnston, E. B. (1991). Systematic distortions of shape from stereopsis. *Vision Research, 31*(7–8), 1351–1360.

Julesz, B. (1960). Binocular depth perception of computer-generated patterns. *Bell Labs Technical Journal, 39,* 1125–1162.

Koenderink, J. J., van Doorn, A. J., & Todd, J. T. (2009). Wide distribution of external local sign in the normal population. *Psychological Research, 73*(1), 14–22.

Julesz, B. (1995). *Dialogues on perception*. Cambridge, MA: MIT Press.

Koenderink, J. J. (2015). PPP. *Perception, 44,* 473–476.

Koenderink, J. J., van Doorn, A., & Wagemans, J. (2011). Depth. *i-Perception, 2,* 541–564.

Koenderink, J. J., Wijntjes, M. W. A., & van Doorn, A. J. (2013). Zograscopic viewing. *i-Perception, 4*(3), 192–206.

Kotulak, J. C., & Schor, C. M. (1986). The accommodative response to sub-threshold blur and to perceptual fading during the Troxler phenomenon. *Perception, 15*(1), 7–15.

Kruger, P. B. (2009). Aberrations of the eye—Crude flaws or ecological design? *Journal of Optometry, 2*(4), 162–164.

Kruger, P. B., Mathews, S., Katz, M., Aggarwala, K. R., & Nowbotsing, S. (1997). Accommodation without feedback suggests directional signals specify ocular focus. *Vision Research, 37*(18), 2511–2526.

Kruger, P. B., Lopez-Gil, N., & Stark, L. R. (2001). Ocular accommodation and the Stiles-Crawford Effect: Theory and a case study. *Ophthalmic and Physiological Optics, 21,* 339–351.

Kruger, P. B., Stark, L. R., & Nguyen, H. N. (2004). Small foveal targets for studies of accommodation and the Stiles-Crawford effect. *Vision Research, 44*(24), 2757–2767.

Langer, M. S., & Siciliano, R. A. (2015). Are blur and disparity complementary cues to depth? *Vision Research, 107,* 15–21.

Liu, S., Hua, H., & Cheng, D. (2010). A novel prototype for an optical see-through head-mounted display with addressable focus cues. *IEEE Transactions on Visualization and Computer Graphics, 16*(3), 381–393.

López-Gil, N., Rucker, F. J., Stark, L. R., Badar, M., Borgovan, T., Burke, S., et al. (2007). Effect of third-order aberrations on dynamic accommodation. *Vision Research, 47*(6), 755–765.

López-Gil, N., Jaskulski, M. T., Vargas-Martin, F., & Kruger, P. B. (2016). Retinal blood vessels may be used to detect the sign of defocus. *ARVO Abstract.*

Maiello, G., Chessa, M., Solari, F., & Bex, P. J. (2015). The (in)effectiveness of simulated blur for depth perception in naturalistic images. *PLoS One,* 8th October 2015.

Marshall, J., Burbeck, C., Ariely, D., Rolland, J., & Martin, K. (1996). Occlusion edge blur: A cue to relative visual depth. *Journal of the Optical Society of America A, 13,* 681–688.

Mather, G. (1996). Image blur as a pictorial depth cue. *Proceedings of the Royal Society B, 263,* 169–172.

Mather, G. (1997). The use of image blur as a depth cue. *Perception, 26,* 1147–1158.

Mather, G., & Smith, D. R. R. (2000). Depth cue integration: Stereopsis and image blur. *Vision Research, 40,* 3501–3506.

Mather, G., & Smith, D. R. R. (2002). Blur discrimination and its relation to blur-mediated depth perception. *Perception, 31,* 1211–1219.

Metlapally, S., Tong, J. L., Tahir, H. J., & Schor, C. M. (2014). The impact of higher-order aberrations on the strength of directional signals produced by accommodative microfluctuations. *Journal of Vision, 14*(12), 19.

Metlapally, S., Tong, J. L., Tahir, H. J., & Schor, C. M. (2016). Potential role for microfluctuations as a temporal directional cue to accommodation. *Journal of Vision, 16*(6), 19.

Metzger, W. (1953). Gesetze des Sehens. Frankfurt: Waldemar Kramer.

Mon-Williams, M., & Tresilian, J. R. (1999). Some recent studies on the extraretinal contribution to distance perception. *Perception, 26,* 167–181.

Mon-Williams, M., & Tresilian, J. R. (2000). Ordinal depth information from accommodation? *Ergonomics, 43,* 391–404.

Mon-Williams, M., Tresilian, J. R., & Roberts, A. (2000). Vergence provides veridical depth perception from horizontal retinal image disparities. *Experimental Brain Research, 133*(3), 407–413.

Morgan, M. J. (1989). Vision of solid objects. *Nature, 339*(6220), 101–103.

Morrison, J. D., & Whiteside, T. C. D. (1984). Binocular cues in the perception of distance of a point source of light. *Perception, 13*, 555–566.

Morrison, K. A., Seidel, D., Strang, N. C., & Gray, L. S. (2010). The effect of proximity on open-loop accommodation responses measured with pinholes. *Ophthalmic and Physiological Optics, 30*(4), 365–370.

Narain, R., Albert, R. A., Bulbul, A., Ward, G. J., Banks, M. S., & O'Brien, J. F. (2015). Optimal presentation of imagery with focus cues on multi-plane displays. *ACM Transactions on Graphics, 34*(4), 59.

Navarro, R. (2009a). The optical design of the human eye: A critical review. *Journal of Optometry, 2*(1), 3–18.

Navarro, R. (2009b). Darwin and the eye. *Journal of Optometry, 2*(2), 59.

Nguyen, V. A., Howard, I. P., & Allison, R. S. (2005). Detection of the depth order of defocused images. *Vision Research, 45*(8), 1003–1011.

O'Shea, R., Govan, D., & Sekuler, R. (1997). Blur and contrast as pictorial depth cues. *Perception, 26*, 599–612.

Otero Molins, C., Aldaba, M., Martínez-Navarro, B., & Pujol, J. (2016). Peripheral depth cues for accommodation stimulation. *ARVO Abstract.*

Palmer, S., & Brooks, J. (2008). Edge-region grouping in figure-ground organization and depth perception. *Journal of Experimental Psychology: Human Perception and Performance, 34*, 1353–1371.

Pentland, A. P. (1987). A new sense for depth of field. *IEEE Transactions on Pattern Analysis and Machine Intelligence, 9*(4), 523–531.

Polans, J., Jaeken, B., McNabb, R. P., Artal, P., & Izatt, J. A. (2015). Wide-field optical model of the human eye with asymmetrically tilted and decentered lens that reproduces measured ocular aberrations. *Optica, 2*(2), 124–134.

Potmesil, M., & Chakravarty, I. (1982). Synthetic image generation with a lens and aperture camera model. *ACM Transactions on Graphics, 1*(2), 85–108.

Rogers, B. (2007). Optic arrays and celestial spheres. *Perception, 36*, 1269–1273.

Rogers, B. (2009). 'Three cheers for retinal images'—A reply. *Perception, 38*, 159–161.

Rogers, B. J., & Bradshaw, M. F. (1995). Disparity scaling and the perception of frontoparallel surfaces. *Perception, 24*, 155–179.

Rogers, B. J., & Brecher, K. (2007). Straight lines, 'uncurved lines', and Helmholtz's 'great circles on the celestial sphere'. *Perception, 36*(9), 1275–1289.

Sebastian, S., Burge, J., & Geisler, W. S. (2015). Defocus blur discrimination in natural images with natural optics. *Journal of Vision, 15*(5), 16.

Sprague, W. W., Cooper, E. A., Reissier, S., Yellapragada, B., & Banks, M. S. (2016). The natural statistics of blur. *Journal of Vision, 16*(10), 23, 1–27.

Stark, L. R., Kruger, P. B., Rucker, F. J., Swanson, W. H., Schmidt, N., Caitlin, H., et al. (2009). Potential signal to accommodation from the Stiles-Crawford effect and ocular monochromatic aberrations. *Journal of Modern Optics, 56*(20), 2203–2216.

Taylor, C. P., & Bex, P. J. (2015). On the number of perceivable blur levels in naturalistic images. *Vision Research, 115,* 142–150.

Thibos, L. N., & Liu, T. (2016). *Using eye models to describe ocular wavefront aberrations.* Wavefront Congress, San Francisco wavefrontcongress.org/Thibos_EyeModels.pdf. Accessed 21 Apr 2016.

Tresilian, J. R., Mon-Williams, M., & Kelly, B. M. (1999). Increasing confidence in vergence as a cue to distance. *Proceedings of the Royal Society B, 266*(1414), 39–44.

Viguier, A., Clément, G., & Trotter, Y. (2001). Distance perception within near visual space. *Perception,30,* 115–124.

Vishwanath, D. (2012a). The utility of defocus blur in binocular depth perception. *i-Perception, 3*(8), 541–546.

Vishwanath, D. (2012b). Counter response to Held, Cooper, & Banks. Originally published alongside Vishwanath (2012a) online, currently unavailable.

Vishwanath, D. (2014). Towards a new theory of stereopsis. *Psychological Review, 121*(2), 151–178.

Vishwanath, D., & Blaser, E. (2010). Retinal blur and the perception of egocentric distance. *Journal of Vision, 10*(10), 26.

Vishwanath, D., & Hibbard, P. B. (2013). Seeing in 3D with just one eye: Stereopsis without binocular vision. *Psychological Science, 24*(9), 1673–1685.

Volcic, R., Vishwanath, D., & Domini, F. (2014). Reaching into pictorial spaces. *Proceedings of SPIE, 9014: Human Vision and Electronic Imaging XIX.*

von Helmholtz, H. (1866). Physiological optics, vol. 3. In J. P. C. Southall (Trans. & ed.) (1925). *Treatise on physiological optics.* New York: Dover.

Wahlberg, M., Lindskoog Pettersson, A., Rosén, R., Nilsson, M., Unsbo, P., & Brautaset, R. (2011). Clinical importance of spherical and chromatic aberration on the accommodative response in contact lens wear. *Journal of Modern Optics, 58*(19–20), 1696–1702.

Wang, B., & Ciuffreda, K. J. (2006). Depth-of-focus of the human eye: Theory and clinical implications. *Survey of Ophthalmology, 51*(1), 75–85.

Wang, B., Ciuffreda, K. J., & Irish, T. (2006). Equiblur zones at the fovea and near retinal periphery. *Vision Research, 46,* 3690–3698.

Wang, Y., Kruger, P. B., Li, J. S., Lin, P. L., & Stark, L. R. (2011). Accommodation to wavefront vergence and chromatic aberration. *Optometry and Vision Science, 88,* 593–600.

Watt, R. J., & Morgan, M. (1983). The recognition and representation of edge blur: Evidence for spatial primitives in human vision. *Vision Research, 23,* 1465–1477.

Watt, S. J., Akeley, K., Girshick, A. R., & Banks, M. S. (2005). Achieving near-correct focus cues in a 3-D display using multiple image planes. *Proceedings of SPIE: Human Vision and Electronic Imaging (IS&T/SPIE Paper Number 5666–53).*

Wilson, B. J., Decker, K. E., & Roorda, A. (2002). Monochromatic aberrations provide an odd-error cue to focus direction. *Journal of the Optical Society of America A, 19*(5), 833–839.

Yang, Z., & Purves, D. (2003). A statistical explanation of visual space. *Nature Neuroscience,* 6, 632–640.

Zannoli, M., Love, G. D., Narain, R., & Banks, M. S. (2016). Blur and the Perception of Depth at Occlusions. *Journal of Vision, 16*(6), 17.

Erratum to: The Perception and Cognition of Visual Space

Erratum to:
P. Linton, *The Perception and Cognition of Visual Space*
https://doi.org/10.1007/978-3-319-66293-0

In the original version of the book, the post-publication corrections from author have been incorporated. The erratum book has been updated with the changes.

The updated online version of the book can be found at
https://doi.org/10.1007/978-3-319-66293-0

E1

INDEX

0-9
2D-plus, 11
8K Super Hi-Vision, 109

A
Absolute depth, 77, 83, 134
Accommodation, 2, 3, 9, 10, 15, 104,
   117–120, 122–133, 136, 140,
   143–146
Affine depth, 3, 12
Alcohol, 64
Amblyopia, 88
Ames room, 18, 20
Anxiety, 64
Aperture, 97, 105, 106, 120, 121,
   142, 148, 149
Astigmatism, 122, 141, 146
Autism, 65

B
Badal lens, 128, 131, 132
Bayesian, 13, 22, 23, 36, 37
Beholder's share, 3
Bias, 13, 14, 24, 36, 54, 55, 63, 121

Binocular depth inversion illusion
   (BDII), 63, 64, 75
Binocular disparity, 2, 3, 5–8, 10, 11,
   16, 23, 35, 36, 45–48, 50, 56–
   58, 63, 65, 73–76, 81–83, 85–88,
   93, 95, 100, 109, 134–136
Bipolar, 64
Blur, 9, 11, 39, 79, 80, 104, 118–123,
   129–132, 136–149

C
Camouflage, 6, 109, 110
Cannabis, 64
Carpentered world, 20
Change-blindness, 45
Child development, 64, 65
Chromatic aberration, 119, 120, 132,
   141–145
Cognitive bias, 54
Cognitively impenetrable/impenetra-
   bility, 12
Cognitive phenomenology, 97, 98
Cognitive psychology, 3, 6, 137
Cognitive revolution, 3–5, 7, 10
Coma, 122, 146

© The Editor(s) (if applicable) and The Author(s) 2017
P. Linton, *The Perception and Cognition of Visual Space*,
DOI 10.1007/978-3-319-66293-0